MCBU
Molecular and Cell Biology Updates

Series Editors:

Prof. Dr. Angelo Azzi
Institut für Biochemie
und Molekularbiologie
Bühlstr. 28
CH–3012 Bern
Switzerland

Prof. Dr. Lester Packer
Dept. of Molecular
and Cell Biology
251 Life Science Addition
Membrane Bioenergetics Group
Berkeley, CA 94720
USA

Signal Transduction in Plants

Edited by   P. Aducci

Birkhäuser Verlag
Basel · Boston · Berlin

Volume editors' address:

Prof. P. Aducci
University of Rome "Tor Vergata"
Department of Biology
Via della Ricerca Scientifica
I-00133 Rome
Italy

A CIP catalogue record for this book is available
from the Library of Congress, Washington D.C., USA

Deutsche Bibliothek Cataloging-in-Publication Data

Signal transduction in plants / ed. by P. Aducci.
 - Basel ; Boston ; Berlin :
Birkhäuser, 1997
(Molecular and cell biology updates)
ISBN 3-7643-5307-4 (Basel ...)
ISBN 0-8176-5307-4 (Boston)
NE: Aducci, Patrizia [Hrsg.]

© 1997 Birkhäuser Verlag, PO Box 133, CH-4010 Basel, Switzerland
Printed on acid-free paper produced from chlorine-free pulp.
Printed in Germany

ISBN 3-7643-5307-4
ISBN 0-8176-5307-4

9 8 7 6 5 4 3 2 1

Table of Contents

Preface

Increasing interest has been emerging in the last decade in the field of signal recognition and transduction. This is particularly true for animal systems where an impressive amount of literature is appearing and where many important pathways have been clarified at a molecular level. In the elucidation of the functions of single components of a given pathway, gene cloning has played a major role and opened the field to the genetic engineering of these complex systems. At variance with this situation, plant systems are less well elucidated, even if in recent years exciting research developments have been initiated especially with the view toward the most promising role of plants in biotechnology.

Recent studies have elucidated some of the events involved in the perception of the plant hormone signals and some steps concerning its transduction. Only for three of the five hormones in plants, namely auxin, ethylene and cytokinins, have specific receptors been isolated. The use of classical molecular approaches, together with the more recently isolated mutants, have produced crucial information on receptors and shed light on possible transduction pathways. As in the case of red light, more than one pathway can be triggered by one specific signal.

Many systems involved in animal signaling are now shown to be present also in plants, and in view of the fast progress in this area, it will be possible in the near future to fully describe the content of the "black boxes" in the reaction chain specifically triggered by a signal.

These have been some of the reasons that prompted us to prepare the present book. It reflects the current "state of the art" in the field of signal transduction in plants. The contributing authors are experts in different areas of plant physiology and plant molecular biology and have used different approaches to study the recognition and transduction of different signals specific for plant organisms. These can be chemical or physical in nature and some examples will be presented on the molecular mechanisms by which hormones, toxins or light can regulate plant growth, differentiation and morphogenesis. Moreover, plant-pathogen interactions are now beginning to be clarified and some progress on their molecular basis is being described.

This book, for its multidisciplinary approach, for the different technologies used and for the emerging interesting peculiarities of the plant world should be of interest both to experts in the field and to a wider audience.

I would like to thank my coworkers Vincenzo Fogliano, Maria Rosaria Fullone, Mauro Marra and Silvia Bacocco for their help in organizing the material for the book and the staff of Birkhäuser for their collaboration.

Patrizia Aducci

Signal Transduction in Plants
P. Aducci (ed.)
© 1997 Birkhäuser Verlag Basel/Switzerland

Roles of ion channels in initiation of signal transduction in higher plants

J.M. Ward and J.I. Schroeder

Department of Biology and Center for Molecular Genetics, University of California San Diego, La Jolla, CA 92093-0116, USA

Introduction

Recently, progress has been made in identifying initial signal reception mechanisms and early events in signaling cascades in higher plants. Ion channels, along with intracellular signaling proteins and second messengers, are critical components mediating early events in higher plant signal transduction. Ion channel-mediated signal transduction in higher plants has notable differences from signaling mechanisms in animal systems. Of the many types of ion channels found in higher plants, there are indications that anion channels, along with Ca^{2+} channels, play critical and rate-limiting roles in the mediation of early events of signal transduction. We have now begun to obtain the first insights into the modes of regulation, the membrane localization, and in the case of K^+ channels, the molecular structure of higher plant ion channels.

In this chapter we focus mainly on novel findings concerning the function and regulation of anion and Ca^{2+} channels and outline testable models of their involvement in signal transduction. Our objective is to summarize these findings and to point out the many open questions involving early events in plant signal transduction. To illustrate the functions of higher plant ion channels in the initiation of signaling cascades, in the first section we discuss the relatively better understood molecular mechanisms of abscisic acid (ABA)-induced stomatal closing with a special focus on new and emerging concepts. In the second section, we address Ca^{2+}-dependent and Ca^{2+}-independent signaling processes in plants and analyze certain putative parallels between initial guard cell signaling and both the initiation of defense responses and phytochrome-induced signaling.

Abscisic acid-initiated signal transduction in guard cells

Since the initial patch-clamp studies of guard cells showed a role for K^+ channels in mediating stomatal movements (Schroeder et al., 1984), guard cells have become a model system for understanding early signal transduction events in plants. A variety of signals, including hormones, light, humidity, and water status, influence stomatal aperture, allowing plants to balance CO_2 uptake and water loss under diverse environmental conditions. The hormone ABA is synthesized in response to drought and induces a cascade of signaling events in guard cells resulting in stomatal closing (Mansfield et al., 1990). Recent research has led to the identification of several early events in ABA signaling, providing a potent system to analyze the intermediate steps in this transduction cascade.

The events that directly follow ABA receptor activation have proven difficult to identify, primarily because of a general lack of information concerning the cellular location and structure of the ABA receptor. Experiments involving microinjection of ABA have indicated a requirement for extracellular ABA receptors (Anderson et al., 1994; Gilroy and Jones, 1994). However, the pH dependence of ABA inhibition of stomatal opening (Anderson et al., 1994, and references therein), as well as the effects of the release of caged ABA within guard cells (Allan et al., 1994) and the effects of microinjected ABA on stomatal closing in the presence of 1 µM extracellular ABA (Schwartz et al., 1994) argues for a role of cytosolic ABA in guard cell signaling. In addition, ABA-induced $^{86}Rb^+$ efflux from guard cells appears to depend on the concentrations of

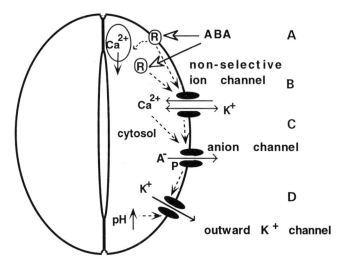

Figure 1. Simplified model for the coordinated regulation of plasma membrane ion channels required for ABA-induced stomatal closing. (A) Activation of abscisic acid receptors leads to Ca^{2+}-dependent and Ca^{2+}-independent signaling events that regulate ion channel activity. (B) The rapid activation of nonselective ion channels causes membrane depolarization and allows Ca^{2+} influx from the extracellular space. The increase in cytosolic Ca^{2+} and the production of other signaling intermediates trigger further Ca^{2+} release from intracellular stores. (C) Membrane depolarization, elevated cytosolic Ca^{2+}, and protein phosphorylation events activate anion channels, which mediate anion release and long-term membrane depolarization. (D) K^+ efflux through voltage dependent outward K^+ channels is driven by membrane depolarization and enhanced by ABA-induced cytosolic alkalinization.

both intracellular and extracellular ABA (MacRobbie, 1995). A model that may account for these circumstantial observations has been proposed in which both extracellular and intracellular sites for ABA reception exist (Anderson et al., 1994; MacRobbie, 1995). A better understanding of the mechanism of ABA perception and the initial targets of activated receptors requires further cellular and molecular identification of the ABA receptors.

One of the earliest known responses of guard cells to ABA is an increase in cytosolic Ca^{2+} (McAinsh et al., 1990; Schroeder and Hagiwara, 1990). Increases in cytosolic Ca^{2+} have been demonstrated to be sufficient to induce stomatal closure (Gilroy et al., 1990). The activation of non-selective, Ca^{2+} permeable ion channels in the plasma membrane that allow Ca^{2+} influx occurs within 2 seconds of ABA exposure (Schroeder and Hagiwara, 1990), indicating a close coupling between ABA receptors and these ion channels. The oscillatory activity of Ca^{2+}-permeable ion channels indicates a close but not direct coupling as depicted in Figure 1 (Schroeder and Hagiwara, 1990). Although it has been suggested that Ca^{2+} influx from the extracellular space is required for stomatal closing (DeSilva et al., 1985; Schwartz et al., 1988), recent work has indicated that Ca^{2+} is released in parallel from intracellular stores: injection of inositol 1,4,5-trisphosphate (InsP3), which is thought to trigger Ca^{2+} release from internal stores in guard cells, produces stomatal closing (Gilroy et al., 1990). In fact, evidence indicates a possible increase in InsP3 levels in guard cells in response to ABA (MacRobbie, 1992; Coté and Crain, 1994). Cytosolic Ca^{2+} (Schroeder and Hagiwara, 1989) and InsP3 (Blatt et al., 1990) also reversibly inhibit inward K^+ channels. Furthermore, the inhibition of inward-rectifying K^+ channels in guard cells following ABA application is not dependent on extracellular Ca^{2+} but is abolished by intracellular application of Ca^{2+} buffers. These data suggest that ABA-induced Ca^{2+} release from intracellular stores can mediate K^+ channel inhibition, while not excluding the occurrence of Ca^{2+} influx at the plasma membrane (Lemtiri-Chlieh and MacRobbie, 1994). In support of these findings suggesting multiple cellular sources of Ca^{2+} influx into the cytosol, imaging studies have shown that ABA induces transient cytosolic Ca^{2+} increases in the vicinity of both the plasma membrane and the vacuole in guard cells (McAinsh et al., 1992). More recently, oscillations in cytoplasmic free Ca^{2+} in guard cells, which result in stomatal closure, have been directly demonstrated to depend on both Ca^{2+} influx and intracellular Ca^{2+} release (McAinsh et al., 1995). These data, as well as other findings on Ca^{2+}-independent signaling (Allan et al., 1994, discussed later in this chapter), indicate that guard cells provide a good system to analyze parallel signaling mechanisms. It is becoming evident that multiple Ca^{2+} channels exist in guard cells and the immediate questions of signal specificity and coupling mechanisms require further analysis. For example, it has not yet been directly demonstrated whether InsP3 induces Ca^{2+} influx from intracellular stores in guard cells or from the extracellular space.

Regulation of K^+ channels and H^+ pumps during guard cell signaling

Increases in cytosolic Ca^{2+} inhibit cellular components that are important in controlling stomatal opening. Plasma membrane H^+-ATPase activity that drives stomatal opening is inhibited almost completely when cytosolic Ca^{2+} is elevated to 1 µM (Kinoshita et al., 1995). The inhibition of H^+ pumping would inhibit stomatal opening, but alone is unlikely sufficient to induce stomatal closure, as described below. Proton pump inhibition is not sufficient to generate the membrane depolarization positive of the K^+ equilibrium potential required for stomatal closing. Inward-rectifying K^+ channels in the guard cell plasma membrane are also inhibited by elevated cytosolic Ca^{2+} (Schroeder and Hagiwara, 1989; Blatt et al., 1990; Fairley-Grenot and Assmann, 1991; Lemtiri-Chlieh and MacRobbie, 1994; Kelly et al., 1995). Recent work demonstrated that strongly elevated cytosolic Ca^{2+} levels in the range of 1 to 1.3 µM are required to significantly inhibit inward-rectifying K^+ channels in guard cells (Kelly et al., 1995). This regulation of K^+ influx is considered to be of secondary importance for the stomatal *closing* mechanisms discussed here because K^+ uptake channels provide a major pathway for K^+ accumulation during stomatal *opening* (Schroeder et al., 1987; Schroeder, 1988; Schroeder and Hagiwara, 1989; Thiel et al., 1992) and have not been proposed to contribute to stomatal closing. Recent findings on second messenger regulation of these K^+ uptake channels are reviewed elsewhere (Assmann, 1993).

Potassium channel currents in guard cells are much larger than required for physiological K^+ fluxes during stomatal movements. Thus, inward K^+ channels in guard cells were initially characterized as sufficient but not rate limiting for stomatal opening. It was stated that "residual K^+ currents measured after reduction by 10 mM Ba^{2+} (90% block) were estimated still to be sufficient to allow stomatal movements in physiologically observed periods" (Schroeder et al., 1987 p. 4112). Recent data have shown that the rate of stomatal opening is only decreased when inward K^+ channels are blocked by 90% (Kelly et al., 1995). Therefore, stomatal opening is likely to be controlled by the rate-limiting plasma membrane H^+ pumps which are tightly regulated in guard cells (Kinoshita et al., 1995).

Stomatal closing is driven by the reduction in guard cell turgor, which requires the efflux of large amounts of K^+ and anions and a parallel conversion of malate to starch (Raschke, 1979; MacRobbie, 1981; Outlaw, 1983). Outward-rectifying K^+ channels were identified and characterized as a pathway for K^+ efflux during stomatal closing (Schroeder et al., 1984; Schroeder et al., 1987; Schroeder, 1988; Blatt, 1990; Blatt and Armstrong, 1993). Detailed studies have shown that K^+ efflux can be mediated by outward-rectifying K^+ channels in the plasma membrane that are activated by membrane depolarization (Fig. 1) (Schroeder et al., 1987; Schroeder, 1988; Blatt and Armstrong, 1993). ABA enhances outward K^+ channel currents (Blatt, 1990) *via* an ABA-induced alkalization of the guard cell cytosol (Irving et al., 1992; Blatt and Armstrong, 1993). However, enhancement of K^+ channel activity alone polarizes the membrane potential to the K^+ equilibrium potential and therefore does not lead to the sustained K^+ efflux required for stomatal closing. The average physiological rate of K^+ efflux during stomatal closing in *Vicia faba* corresponds to a K^+ current density of approximately 4.2 μA cm^{-2}. Guard cell K^+ efflux channel activity, both before and after ABA exposure is sufficiently large ([>10 $\mu A \cdot cm^{-2}$ at 0 mV] Schroeder et al., 1987; Blatt, 1990; Armstrong and Blatt, 1993) to provide the required physiological K^+ fluxes during stomatal closing (see Schroeder et al., 1987). We therefore propose that outward K^+ channel modulation of stomatal by ABA can accelerate stomatal closing, but that initiation of stomatal closing requires a more tightly regulated mechanism. Therefore a mechanism is required that is more strongly up- and down-regulated and can depolarize guard cells to drive the net K^+ release required for stomatal closing.

Guard cell anion channels control ion efflux during stomatal closing

Abscisic acid triggers long-term plasma membrane depolarization in guard cells (Kasamo et al., 1981; Ishikawa et al., 1983; Thiel et al., 1992). Patch-clamp studies have led to the identification of anion channel currents in guard cells, which have been suggested to provide a mechanism for the required depolarization during stomatal closure (Schroeder and Hagiwara, 1989). These plasma membrane anion channels allow Cl^- and malate efflux (Keller et al., 1989; Schroeder and Hagiwara, 1989; Schmidt and Schroeder, 1994) and the resulting depolarization serves to both activate K^+ efflux channels and drive K^+ release, as illustrated in Figure 1. Because of their roles in mediating anion efflux (directly) and in regulating K^+ efflux (indirectly), anion channels have been suggested to function as a central control mechanism for stomatal closure (Schroeder and Hagiwara, 1989). Furthermore, anion channels are activated at elevated cytosolic Ca^{2+} levels (Schroeder and Hagiwara, 1989; Hedrich et al., 1990) and may therefore provide a link in the ABA signaling pathway that couples Ca^{2+} increases with the anion and K^+ efflux required for stomatal closure. In general terms, a parallel to the classically described action potentials in algae can be found here, in which Ca^{2+} influx causes activation of Cl^- currents (Williamson and Ashley, 1982; Lunevsky et al., 1983; Shiina and Tazawa, 1988; Mimura and Shimmen, 1994; Shimmen et al., 1994). It should be noted that K^+ efflux channels in guard cells appear not to be activated by physiological increases in cytosolic Ca^{2+} concentration (for example Lemtiri-Chlieh and MacRobbie, 1994). However, very high cytosolic Ca^{2+} concentrations (buffered to >10 μM) appear to inhibit K^+ efflux channels (W.B. Kelly and J.I. Schroeder, unpublished data).

Two modes of anion channel activity have been identified in guard cells: slow (S-type) (Schroeder and Hagiwara, 1989; Schroeder and Keller, 1992) and rapid (R-type) (Keller et al., 1989; Hedrich et al., 1990). Figure 2 illustrates the general predicted time-courses of the membrane depolarizations produced by activation of S- and R-type anion channel activity. The slow and sustained activation of S-type anion channels would produce a long-term depolarization (that is, in the range of several minutes or longer) (Schroeder and Hagiwara, 1989; Schroeder and Keller, 1992; Linder and Raschke, 1992) whereas the rapid activation and inactivation of R-type anion channels would produce a short, transient depolarization (Hedrich et al., 1990). Despite the extreme (10^3-fold) differences in kinetics (Schroeder and Keller, 1992), these two anion channel modes may share common structural components (Schroeder et al., 1993; Zimmermann et al., 1994; for review, see Schroeder, 1995). The most compelling support for this hypothesis has been obtained in studies of tobacco, where a shift from R- to S-type anion current properties occurs when ATP is removed from the cytosol (Zimmermann et al., 1994). This transition is inhibited by phosphatase inhibitors suggesting ATP-dependent phosphorylation events.

Because S-type anion channels can produce the sustained depolarization that is required for driving K^+ channel-mediated K^+ efflux (Figs 1 and 2), they have been proposed to provide a central mechanism for controlling stomatal closure (Schroeder and Hagiwara, 1989; Schroeder and Keller, 1992). Pharmacological studies have provided correlative support for this hypothesis: when both S- and R-type anion channels are blocked by the inhibitors NPPB (5-nitro-2,3-phenylpropyllaminobenzoic acid) or A9C (anthracene-9-carboxylate), ABA and malate-induced stomatal closing is inhibited completely (Schroeder et al., 1993; Schwartz et al., 1995). However,

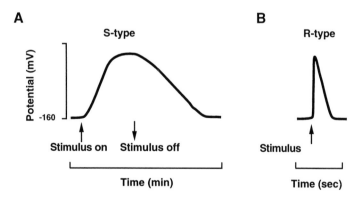

Figure 2. Simplified drawing of time-courses of plasma membrane depolarizations in response to the activation of slow and rapid anion channel activities. (A) Slow (S-type) anion channel. (B) Rapid (R-type) anion channel. Note the different time scales in A and B.

DIDS (4,4'-diiosthiocyanatostilbene-2,2'-disulfonic acid), a potent blocker of R-type anion channels ($K_i = 0.2 \, \mu$M) (Marten et al., 1993) that does not inhibit S-type anion channels, had no significant effect on ABA and malate-induced stomatal closing (Schroeder et al., 1993). Furthermore, the concentration dependence of NPPB and A9C block of anion channels correlates to the NPPB and A9C concentration dependence for inhibition of stomatal closing (Schroeder et al., 1993; Schwartz et al., 1995). Although biophysical, cell biological, and pharmacological data support a rate-limiting role for S-type anion channels in stomatal closure, they do not exclude an additional role for R-type anion channels in this process.

The effects of slow anion channel blockers on ABA-induced stomatal closing are very potent. For example, we have observed that the inhibition of ABA-induced stomatal closing by anion channel blockers is more striking than the effects of voltage-dependent calcium channels blockers. In addition, the ability of extracellular malate, which has been proposed to directly activate anion channels (Hedrich et al., 1994), to produce partial stomatal closing is also very consistent. The strong and consistent effects of these anion channel blockers and activators provide tools to explore the regulation of stomatal aperture. We note however, that pharmacological tools do not provide definitive proof for the model in Figure 1 and that future molecular approaches will allow a more detailed analysis. For reasons that are unclear, we have not been able to reproduce the reported induction of stomatal closing using very low (0.2 mM) malate concentrations (Hedrich et al., 1994), even when using the same *Vicia faba* seed stock (S. Kröbel, Göttingen, Germany), the same growth conditions, and the same solutions used by Hedrich et al. (1994). Instead, we have found consistently, in >60 experiments under various conditions, half-maximum stomatal closing at 20 mM malate, with saturation at ~40 mM malate (Esser et al., 1996). Interestingly, stomatal closing in response to malate is not observed if stomatal apertures are opened very wide (>15 μm for *Vicia faba*), suggesting that anion channels may be completely downregulated during wide stomatal opening and resistant to modulation by malate (Esser et al., 1996).

Strong regulation of slow anion channels by Ca^{2+} and phosphorylation events

Recent studies have suggested that slow anion channel activation is required to drive stomatal closing and strong inactivation of these ion channels is necessary during stomatal opening. Therefore, it is important to understand the mechanism of slow anion channel regulation. S-type anion channels in guard cells have been shown to be activated by both increases in cytoplasmic Ca^{2+} and membrane depolarization (Schroeder and Hagiwara, 1989; Schroeder and Keller, 1992). Parallel pathways that induce both intracellular Ca^{2+} release and Ca^{2+} influx in response to ABA may provide mechanisms to activate anion channels in guard cells (Fig. 1) (McAinsh et al., 1990; Schroeder and Hagiwara, 1990; Cosgrove and Hedrich, 1991; Allen and Sanders, 1994; Ward and Schroeder, 1994). However, the positive regulatory effect of cytosolic Ca^{2+} on slow anion channels may be indirect (Schroeder and Hagiwara, 1989). Furthermore, recent studies suggest that ABA can also stimulate stomatal closing *via a Ca^{2+}-independent* signaling pathway (Allan et al., 1994). Because S-type anion channels are rate-limiting for stomatal closure, it is likely that they are integral components of both Ca^{2+}-dependent and Ca^{2+}-independent pathways (Allan et al., 1994), which may converge to activate these ion channels.

Slow anion channels have a weak voltage dependence (Schroeder and Hagiwara, 1989; Schroeder and Keller, 1992), therefore other regulatory mechanisms must be responsible for the complete downregulation during stomatal opening and strong activation during stomatal closing. In patch-clamp experiments, slow anion channel activity in *Vicia faba* is rapidly abolished when ATP is excluded from the cytosolic solution or by the kinase inhibitors K-252a or H7 (Schmidt et al., 1995). These data, as well as the inability of GTP or nonhydrolyzable analogs of ATP or GTP to replace ATP, indicate that a phosphorylation event activates slow anion channels. Furthermore, the phosphatase inhibitor okadaic acid maintained slow anion channel activity in the absence of cytosolic ATP (Schmidt et al., 1995). Consistent with these findings, ABA-induced stomatal closing was abolished by kinase inhibitors and enhanced by phosphatase inhibitors (Schmidt et al., 1995). Kinase activation and/or okadaic acid-sensitive phosphatase inhibition may therefore be part of the ABA signaling pathway leading to regulation of slow anion channels. This strong regulation of slow anion channels by phosphorylation/dephosphorylation events results in a large dynamic range of anion channel activity required to control stomatal aperture (Schmidt et al., 1995). Further research will be required to determine what mechanisms directly activate anion channels, whether ABA actually activates anion channels, and whether Ca^{2+}-independent signaling pathways produce anion channel activation through these signaling mechanisms.

It is interesting that slow anion channels can be completely inactivated and strongly activated (Schroeder and Hagiwara, 1989; Schmidt et al., 1995) while the dynamic range of K^+ efflux channels is more limited (as discussed above). Complete down-regulation of K^+ efflux channels would, in theory, effectively inhibit stomatal closing according to the model in Figure 1. It is possible that other important functions of K^+ efflux channels (such as membrane potential control) prohibit a stronger regulation of their activity during stomatal movements.

Vacuolar K^+ and Ca^{2+} release channels and stomatal regulation

Whereas detailed studies have led to a model for a cascade of ion channels in the guard cell plasma membrane that mediate stomatal closing (Fig. 1), the mechanisms for parallel K^+, Ca^{2+}, and anion release from guard cell vacuoles are only just beginning to be understood. The guard cell vacuole functions as a storage organelle for solutes that are important for osmoregulation during stomatal movements. The vacuole constitutes >90% of the guard cell volume, and >90% of the K^+ and anions released from guard cells during stomatal closing must first be released from vacuoles into the cytosol (MacRobbie, 1981). Recent studies have revealed the existence of a novel highly selective vacuolar K^+ channel (VK channel) that may provide an important pathway for K^+ release from the vacuole (Ward and Schroeder, 1994). VK channels are not measurably active when cytosolic Ca^{2+} is buffered to low concentrations, but are strongly and rapidly acti-vated when cytosolic Ca^{2+} is increased within the physiological range to ~1 μM. These properties of VK channels are distinct from those of fast vacuolar (FV) ion channels, which are inhibited by increases in cytosolic Ca^{2+} and have been proposed to allow both K^+ and anion conductance (Hedrich and Neher, 1987). Estimates show that the large number of VK channels in guard cell vacuoles may provide a major pathway for K^+ release, as illustrated in Figure 3.

Long-term K^+ release from the vacuole, which is necessary for stomatal closure, would require a mechanism that electrically drives K^+ out of vacuoles, such as the activity of vacuolar H^+ pumps. An increased pumping of protons into the vacuole during VK channel activation could contribute to the ABA-induced cytoplasmic alkalinization in guard cells observed during stomatal closure (Irving et al., 1992). Interestingly, ABA-induced cytosolic alkalinization, in turn enhances K^+ efflux channel activity in the plasma membrane of guard cells (Blatt and Armstrong, 1993), providing a potential mechanism for cross-talk between K^+ efflux mechanisms in the vacuolar and plasma membranes (Figs 1 and 3).

The selective release of K^+ from vacuoles *via* the large VK channel conductance (Ward and Schroeder, 1994) causes a shift in the vacuolar membrane potential to more positive potentials on the cytosolic side (Fig. 3). The positive shift in vacuole membrane potential, resulting from VK channel activity could lead to the activation of the ubiquitous slow vacuolar ion channels (SV channels) (Hedrich and Neher, 1987; Kolb et al., 1987; Colombo et al., 1988; Amodeo et al., 1994). SV channels differ in many respects from VK channels; they are highly voltage-depen-dent, non-selective cation channels as characterized in detail in barley mesophyll cells (Kolb et al., 1987), *Acer* cells (Colombo et al., 1988), and *Allium* guard cell vacuoles (Amodeo et al., 1994). However, like VK channels, SV channels are activated by cytosolic Ca^{2+} (Hedrich and Neher, 1987). Interestingly, SV channels from guard cell and red beet storage tissue were recently deter-

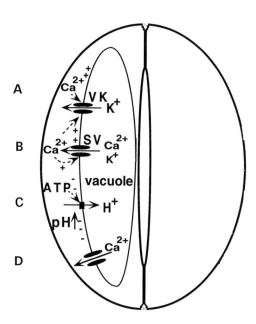

Figure 3. Model for ion channel-mediated K⁺ and Ca²⁺ release from guard cell vacuoles during stomatal closure. (A) Increases in cytoplasmic Ca^{2+} activate selective vacuolar K⁺ (VK) channels in the vacuolar membrane. The resulting K⁺ influx into the cytoplasm causes a positive shift in vacuole membrane potential. (B) The activation of slow vacuolar (SV) non-selective cation channels leads to the release of K⁺ and Ca^{2+} from the vacuole. Increased cytoplasmic Ca^{2+} provides positive feedback to further activate both SV and VK ion channels. (C) Vacuolar proton pump activity is required to drive long-term K⁺ efflux through VK channels and provides a mechanism for cytoplasmic alkalinization observed in response to ABA. (D) Ca^{2+} release channels activated by alkalinization of the vacuolar lumen allow Ca^{2+} release from vacuoles at negative potentials on the cytosolic side of the membrane.

mined to be selective for Ca^{2+} (permeability ratio for Ca^{2+} to K⁺ of $\approx$ 3:1) (Ward and Schroeder, 1994). The single channel conductance of SV channels in guard cell vacuoles is reduced from $\geq$ 70 pS in K⁺ solutions to ~16 pS when Ca^{2+} is the only permeant ion (Ward and Schroeder, 1994), indicating a strong binding site for Ca^{2+} in the channel pore (Hille, 1992). These data indicate that SV channels may contribute to the release of both K⁺ and Ca^{2+} from vacuoles during stomatal closure (Fig. 3) (Ward and Schroeder, 1994).

Because SV channels are not strongly selective, data have been interpreted to presume Cl and glutamate permeation and to assume that they may also function in anion release from vacuoles (Hedrich and Neher, 1987; Schulz-Lessdorf et al., 1994; Schulz-Lessdorf and Hedrich, 1995) which could be physiologically important for stomatal closing. This possibility of significant Cl⁻ permeability through SV channels was directly analyzed in *Vicia faba* guard cells under physiological K⁺ gradients. In patch-clamp experiments using conditions such that the equilibrium potentials for all available cations are equivalent to one another and the Cl⁻

equilibrium potential differs significantly, the contribution of Cl^- permeability to SV channel currents was negligible (Ward et al., 1995). Furthermore, the Ca^{2+} to K^+ permeability ratio of guard cell SV channels was recently confirmed (Allen and Sanders, 1995). These results confirm previous detailed studies of SV channels showing lack of anion permeability (Kolb et al., 1987; Colombo et al., 1988; Amodeo et al., 1994). Anion permeability through cation selective channels is biophysically unlikely (Hille, 1992). ABA-induced anion release from guard cell vacuoles is required during stomatal closing and the mechanism for the necessary anion release therefore remains unknown. Anion release by a passive ion channel mechanism is perhaps less likely because of the recently proposed need for vacuolar H^+ pump activity during stomatal closure which would tend to drive anion accumulation (Fig. 3) (Ward and Schroeder, 1994).

The finding that Ca^{2+}-activated SV channels are Ca^{2+} permeable has led to the suggestion that these ion channels may provide an important mechanism, not only for K^+ release from guard cell vacuoles, but also for vacuolar Ca^{2+}-induced Ca^{2+} release (CICR) (Ward and Schroeder, 1994). SV channels are ubiquitous in vacuoles from diverse plants and tissues and their proposed function in Ca^{2+}-induced Ca^{2+} release could be important for the transduction of a variety of signals. During ABA-induced signaling in guard cells, Ca^{2+} influx through non-selective ion channels in the plasma membrane (Fig. 1) (Schroeder and Hagiwara, 1990; McAinsh et al., 1990) or InsP3-induced Ca^{2+} release (Gilroy et al., 1990) could trigger vacuolar Ca^{2+} release through SV channels.

Multiple vacuolar Ca^{2+} release channels

In addition to SV channels, several distinct Ca^{2+}-permeable ion channels have been identified in vacuoles of sugar beet and guard cells. Voltage-dependent vacuolar Ca^{2+} channels have been identified that display higher open probability at physiologically negative membrane potentials on the cytosolic side of the membrane (Johannes et al., 1992). These Ca^{2+} channels are distinct from SV channels in that they are insensitive to cytoplasmic Ca^{2+}. Similar ion channels have recently been identified in guard cell vacuoles (Allen and Sanders, 1994). These guard cell vacuolar Ca^{2+} release channels are activated by alkaline pH in the vacuolar lumen of guard cells, indicating that specific phases of stomatal movements that produce an alkaline vacuolar pH may lead to Ca^{2+} release. Another class of Ca^{2+}-permeable ion channels that are inhibited rather than activated by micromolar levels of cytosolic Ca^{2+}, have also been found in beet vacuoles (Gelli and Blumwald, 1993). In addition, cyclic ADP-ribose activated channels have recently been identified in sugar beet vacuoles, that may in turn interact with SV channels in mediating Ca^{2+} induced Ca^{2+} release (Allen et al., 1995).

Critical questions for calcium-induced calcium release

The resting potential of the vacuolar membrane is normally in the range of -20 to -50 mV on the cytosolic side of the membrane due to proton pump activity (for review see Sze et al., 1992). SV ion channels usually activate at vacuolar membrane potentials of 0 mV or more positive, and only when cytosolic Ca^{2+} is increased above the resting level (Hedrich and Neher, 1987). One mechanism that may shift the vacuolar membrane potential sufficiently positive to activate SV ion channels in guard cells is Ca^{2+}-activated K^+ flux from the vacuole to the cytoplasm through VK channels (Ward and Schroeder, 1994) (Fig. 3). Because of the high K^+ selectivity and abundance of VK channels, the ability of VK channel activity to shift the vacuolar membrane potential toward 0 mV depends on the K^+ gradient from the vacuole to the cytoplasm. During stomatal closure, K^+ efflux across the guard cell plasma membrane *via* outward-rectifying K^+ channels (Fig. 1) may maintain a downhill K^+ gradient from the vacuole to the cytoplasm that would shift the vacuolar membrane potential to positive voltages on the cytosolic side that, in turn, activate SV channels. Measurements of K^+ in stomata using electron microprobe analysis indicate a possible gradient of K^+ from the vacuole to the cytosol in guard cells (Humble and Raschke, 1971). However, the use of more direct techniques to measure K^+ levels would be required to determine K^+ gradients during stomatal closing. In addition, intracellular modulators including Ca^{2+} (Hedrich and Neher, 1987), calmodulin (Weiser et al., 1991; Bethke and Jones, 1994), cytoplasmic Cl^- (Pantoja et al., 1992), and other factors, may cause a shift in SV channel activation to less positive potentials that could facilitate signal-dependent Ca^{2+}-induced Ca^{2+} release.

SV channel activity at positive potentials produces net cation flux from the cytosol to the vacuole. However, Ca^{2+} release *via* SV channels would necessitate that Ca^{2+} permeates SV channels in the opposite direction, that is from the vacuole into the cytosol. Recent studies in fact indicate that the large ($>10^4$-fold) physiological gradient of Ca^{2+} from the vacuole to the cytoplasm could give rise to vacuolar Ca^{2+} release even when net currents are directed into the vacuole. For example, the studies of Mayer et al. (1987) and Schneggenburger et al. (1993) demonstrate that nonselective cation channels allow physiologically significant "fractional" Ca^{2+} influx even when net currents through these channels are directed outward. Furthermore, even cation channels that have a low Ca^{2+} permeability and therefore show no shifts in reversal potential with extracellular Ca^{2+} changes, can mediate physiologically significant Ca^{2+} influx into the cytoplasm (Keller et al., 1992). In other words, the very large physiological Ca^{2+} gradient into the cytosol and thermodynamic probabilities of ion permeation in non-selective cation channels can result in physiological Ca^{2+} influx even when cation channels conduct net outward currents (Mayer et al., 1987; Schneggenburger et al., 1993) and even through channels that are only minutely permeant to Ca^{2+}. These findings on Ca^{2+} permeation, taken together with the recent demonstration of Ca^{2+} selectivity of SV channels, and the ubiquity of SV channels in plant vacuoles, indicate a potential role of SV channels for physiological Ca^{2+}-induced Ca^{2+} release.

Calcium-independent and cytosolic signaling in guard cells

The recent advances in our understanding of signaling pathways in higher plants underlines the importance of an emerging theme in plant signal transduction: both Ca^{2+}-dependent and Ca^{2+}-independent pathways are often activated by the same physiological stimulus (Neuhaus et al., 1993; Allan et al., 1994; Bowler et al., 1994). In guard cells, ABA-induced signaling appears to proceed through parallel Ca^{2+}-dependent and Ca^{2+}-independent pathways, both of which may be sufficient to cause stomatal closure (MacRobbie, 1992; Allan et al., 1994). The balance between these two pathways is environmentally influenced, because only those *Commelina* plants grown at elevated temperature produce strong measurable increases in guard cell cytosolic Ca^{2+} in response to ABA (Allan et al., 1994). Therefore, activation of Ca^{2+}-permeable ion channels in guard cells may not be required or may be limited during stomatal closure under growth conditions in which Ca^{2+}-independent signaling predominates. This is consistent with reports in which only a fraction of guard cells responded to ABA with an increase in cytosolic Ca^{2+} (Schroeder and Hagiwara, 1990; Gilroy et al., 1991). Because Ca^{2+}-independent signaling pathways may predominate under some growth conditions, signaling intermediates other than Ca^{2+} are expected to have strong regulatory effects on the ion channels involved in the mediation of stomatal closing (Figs 1 and 3) (Schmidt et al., 1995).

Studies on guard cells have led to the identification of networks of ion channels in the plasma membrane and vacuolar membrane that mediate stomatal closing in response to ABA (Figs 1 and 3). It should be noted that the exact sequence of ion channel activation in this network has not been characterized directly (MacRobbie, 1992) but these processes are expected to be regulated in a parallel, rather than in a strictly linear manner. The analysis of direct mechanisms of regulation of ion channels in these cascades will shed light on the intermediate signaling steps in ABA-induced closing, which have remained largely unknown. The ion channel cascades in guard cells, shown in Figures 1 and 3, provide an excellent backbone for analyzing intermediate events in signal-induced stomatal closing. That is, individual ion channel types may be utilized as biochemical probes to identify upstream and downstream signaling components. This approach has been pursued to study inward K^+ channel regulation during stomatal opening (for review see Assmann, 1993). As an example, mutations in the *ABI1* locus of *Arabidopsis* disrupt ABA-induced stomatal closing (Finkelstein and Sommerville, 1990). *ABI1* has recently been cloned and found to encode a protein phosphatase (Leung et al., 1994; Meyer et al., 1994). Ion channels involved in ABA-induced stomatal closure (Figs 1 and 3) present possible targets for the ABI-1 phosphatase. Analysis of the effects of the ABI1 phosphatase on these individual components may reveal the physiological function and the location within the signaling cascade of ABI1 and should allow refinements in our understanding of direct regulation of ion channels during stomatal closure. Transgenic tobacco plants carrying the mutant *abi1-1* gene were found to have a wilty phenotype, reminiscent of the *Arabidopsis abi1* phenotype (Armstrong et al., 1995). In wild-type tobacco the application of ABA induced a two-fold increase in outward K^+ currents and a decrease in inward K^+ currents. However, in the transgenic plants these responses to ABA were abolished, indicating that ABI1 functions upstream of K^+ channel regulation (Armstrong et al., 1995). Regulation of anion channels by ABI1 could not be addressed in transgenic *abi-1-* expressing tobacco since anion currents were approximately 90% inactivated (average current $6 \ \mu A \ cm^{-2}$) even in ABA treated control plants (Armstrong et al., 1995). Further analysis of the intermediate signaling events should provide insights into the unidentified links between plasma membrane and vacuolar membrane signaling discussed here and links to the important metabolic conversion of malate to starch during stomatal closing (Outlaw et al., 1981).

Anion and voltage-dependent calcium channels: Putative mechanisms for the initiation of signal transduction in plant cells

Plasma membrane depolarization in plant cells, involving a shift from the normally negative resting potentials of -140 to -200 mV to more positive potentials, occurs in response to numerous physiological stimuli. In fact, depolarization in plant cells has been reported in response not only to ABA (Kasamo, 1981; Ishikawa et al., 1983; Thiel et al., 1992), but also to auxin (Bates and Goldsmith, 1983), elicitors of plant defense responses (Mathieu et al., 1991; Kuchitsu et al., 1993), phytotoxins (Ullrich and Novacky, 1991), phytochrome (Racusen and Satter, 1975), blue light (Spalding and Cosgrove, 1989) and Nod factors (Ehrhardt et al., 1992). In plants, several mechanisms could cause plasma membrane depolarization including the inhibition of proton pumps; the activation of anion, Ca^{2+}, or non-selective channels; and K^+ channel modulation (for review, see Schroeder and Hedrich, 1989). Recent research indicates putative roles for anion channels and Ca^{2+} channels in some of the above signal-induced depolarizations as originally described for action potentials in algae and as outlined below.

The information supplied to cells following anion channel-mediated depolarization can be separated into distinct components. The amplitude of the depolarization, which is dependent on the number of activated anion channels, will determine which types of voltage-dependent ion channels, such as recently identified voltage-dependent Ca^{2+}-channels (Thuleau et al., 1994) or voltage-dependent K^+ channels, are modulated. The time course of anion channel activation further controls the time course of the depolarizing signal, as depicted in Figure 2. The extreme differences in this regard between S- and R-type anion channels in guard cells and cultured tobacco cells demonstrate the possibility for a broad range in the timing of depolarizations in plant cells (Schroeder and Keller, 1992; Zimmermann et al., 1994). Anion channels in plants have also been proposed to mediate transport of anionic nutrients, and additional types of anion channels, that are activated by hyperpolarization have been described in higher plant cells (for reviews, see Tyerman, 1992; Schroeder, 1995).

The timing of the onset of anion channel activity is driven by positive feedback regulation because both S- and R-type anion channels are activated by depolarization. Thus, anion efflux in response to activation of a few anion channels produces further depolarization. This depolarization can, in turn, lead to rapid signal amplification as additional anion channels respond to increased depolarization. Data from guard cells, cultured tobacco cells and *Arabidopsis* hypocotyles indicate that anion channel proteins may be directly modulated by blue light, auxin and by other organic acids (Marten et al., 1991; Hedrich et al., 1994; Zimmermann et al., 1994; Cho and Spalding, 1996). The return to resting potential depends both on the rate of anion channel closing and on the activity of proton pumps and other ion channels, such as outward-rectifying K^+ channels, that repolarize the membrane. The question remains: why is anion channel activation a common signal transduction component? We propose two possible reasons: (1) the control of voltage dependent Ca^{2+} signaling, and (2) as described in classical studies in algae, signal propagation (Cole and Curtis, 1938; Gaffey and Mullins, 1958; Shina and Tazawa, 1988).

Voltage-dependent Ca^{2+} channels

Numerous pharmacological and cell biological studies have suggested that voltage-dependent Ca^{2+} channels in the plasma membrane are important for initiation of plant responses to environmental, hormonal, and pathogenic signals (Hepler and Wayne, 1985; Leonard and Hepler, 1990). Direct measurements of such channels in plant cells have been recently reported in patch-clamp studies (Thuleau et al., 1994), in studies using protoplasts loaded with Ca^{2+}-sensitive dyes (Ranjeva et al., 1992), in radioactive tracer flux studies using plasma membrane vesicles (Huang et al., 1994; Marshall et al., 1994), and in reconstitution studies (Piñeros and Tester, 1995). These plant Ca^{2+} channels are activated by membrane depolarization, a characteristic typical of voltage-dependent Ca^{2+} channels in other systems. However, the recently identified plant Ca^{2+} channels appear specifically suited for signal transduction in higher plant cells, which typically have strongly negative membrane potentials in the range of −140 to −200 mV. To date, voltage-dependent Ca^{2+} channels found in higher plants activate at potentials more positive than ∼−140 mV (Huang et al., 1994; Thuleau et al., 1994). This threshold potential for Ca^{2+} channel activation is ∼50 to 90 mV more negative than that of Ca^{2+} channels in animal cells.

The finding that voltage-dependent Ca^{2+} channels in higher plant cells show a steep activation in response to physiologically occurring depolarizations (Thuleau et al., 1994) indicates the importance of plasma membrane transporters that produce depolarization. Plasma membrane anion channels offer a likely mechanism for the initiation and control of membrane depolarizations. As depicted in the simplified model shown in Figure 4, cell stimulation may cause anion channel activation. Anion efflux then results in plasma membrane depolarization which in turn triggers the activation of voltage-dependent Ca^{2+} channels which mediate Ca^{2+} influx. This model shows one way in which receptor-mediated depolarization may be coupled to Ca^{2+} signaling and suggests that signal-induced depolarizations may be indicative of voltage-regulated Ca^{2+} signal transduction or of signal propagation (see later discussion). Further work is necessary to determine whether, as in guard cells (Fig. 1), non-selective cation channels may be involved in the early signal transduction preceding anion channel activation in response to other stimuli. It is important to keep in mind that in Characean algae, anion channel activation depends on Ca^{2+} influx at the plasma membrane (Gaffey and Mullins, 1958; Williamson and Ashley, 1982; Lunevsky et al., 1983; Shiina and Tazawa, 1988; Mimura and Shimmen, 1994; Shimmen et

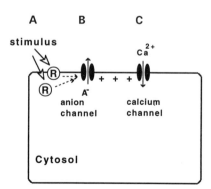

Figure 4. A simplified model for the putative coupling of plasma membrane depolarization and Ca^{2+} influx. (A) Receptor activation triggers initial cytosolic signals. (B) Anion channel activation mediates anion efflux resulting in membrane depolarization. (C) Voltage-dependent Ca^{2+} channels activate in response to depolarization, allowing Ca^{2+} influx. Note that the return to hyperpolarized resting potentials by subsequent outward K^+ channel and H^+ pump activation was omitted here for clarity.

al., 1994). Similar regulation of anion channels may also occur in higher plant cells. In the guard cell plasma membrane, for example, anion channels are also activated by increases in cytoplasmic Ca^{2+} (Schroeder and Hagiwara, 1989) and by extracellular $CaCl_2$ under conditions where the cytosolic Ca^{2+} buffering capacity is low (Hedrich et al., 1990).

Another function of anion channels may be that of electrical signal propagation because anion efflux causes depolarization which in turn can activate adjacent depolarization activated anion channels as has been described for action potential propagation in algae (for review see Tazawa et al., 1987). This electrical signal may also propagate to adjacent cells and play a role in priming tissue for subsequent systemic signals (Wildon et al., 1992). In fact, electrical signal propagation has been shown to be accompanied by anion efflux (Davies, 1987).

Some aspects of plant defense and phytochrome signaling

Early events in plant defense reactions to pathogens involve both plasma membrane depolarization and Ca^{2+} influx: Long-term depolarization, in the range of 30 min or more, has been reported in response to specific elicitors in tobacco (Mathieu et al., 1991) and rice (Kuchitsu et al., 1993). In these systems, depolarization occurs within minutes following elicitor application, preceding the induction of defense-related gene expression. Recent work has demonstrated that hypersensitive response-related cell death, induced by specific avirulent bacterial pathogens in soybean cells, is abolished by anion channel blockers (A. Levine and C. Lamb, personal communication). In parsley cells, a fungal oligopeptide elicitor induces anion efflux and Ca^{2+} influx (Nürnberger et al., 1994). The induction of defense-related genes in parsley cells in response to this elicitor requires extracellular Ca^{2+} (Nürnberger et al., 1994). Furthermore, the elicitor-induced defense gene activation in parsley cells is blocked by anion channel blockers (T. Jabs and D. Scheel, personal communication). In soybean cells, anion channel blockers inhibit β-glucan-induced phytoalexin production and the inducible Ca^{2+} influx (Ebel et al., 1995). All of these data indicate a possible rate-limiting role for anion channel-mediated depolarization in elicitor signaling pathways and would be consistent with the simplified model in Figure 4. Furthermore, these data indicate that receptors specific for fungal or plant-derived elicitors are closely coupled to and activate plasma membrane ion channels important for signal transduction in response to pathogens.

Phytochrome initiates plant responses to red light resulting in changes in plant growth and development. Classic studies have shown that phytochrome stimulation induces plasma membrane depolarization (Racusen and Satter, 1975). In mesophyll cells, white light-induced plasma membrane depolarization is dependent on the external Cl^- concentration, indicating the participation of anion channels (Elzenga et al., 1995). In addition, recent work has shown that the anion channel blocker NPPB abolished white light- and blue light-induced depolarization in pea leaf mesophyll cells and *Arabidopsis* hypoeoxyl cells (J. T. M. Elzenga and E. Van Volkenburgh, personal communication; Cho and Spalding, 1996).

Recent studies have shown that transient changes in cytoplasmic Ca^{2+} are one of the most rapid phytochrome responses. Using etiolated wheat leaf protoplasts, Shacklock et al. (1992) have shown that red light induces protoplast swelling and causes a transient Ca^{2+} increase followed by a rapid reduction in cytosolic Ca^{2+} to below resting levels. Protoplast swelling is also induced by the release of InsP3 in the cytosol indicating that a transient Ca^{2+} increase is sufficient to mimic phytochrome signaling in this system. The importance of cytoplasmic Ca^{2+} in phytochrome signaling has been further demonstrated using a tomato mutant deficient in phytochrome A. Microinjection of Ca^{2+} into single cells of tomato was found to stimulate chloroplast development, partially reversing the mutant phenotype (Neuhaus et al., 1993). Further experiments using this system identified a parallel cGMP-dependent pathway responsible for phytochrome-induced anthocyanin production (Bowler et al., 1994). In these studies, effects of pharmacological activators and inhibitors of GTP binding proteins suggest that G-proteins may be the furthest upstream components following phytochrome activation identified to date in the signaling pathway (Bowler et al., 1994). Biochemical and molecular biological studies will be required to test the roles of putative G-proteins in these processes.

Conclusions

Early events in the responses of plant cells to many physiological stimuli share common features, such as membrane depolarization and elevations in cytosolic Ca^{2+}. Recent results have greatly enhanced our understanding of the early events in the response of guard cells to ABA and the regulation of plasma membrane and vacuolar ion channels required for stomatal closure. Several components identified in guard cells, in particular plasma membrane anion and Ca^{2+} channels and vacuolar Ca^{2+} release channels, may participate in multiple signaling pathways in higher plant cells. Furthermore, other types of ion channels are likely to be identified in guard cells and found to be important in stomatal regulation. For example, voltage-dependent Ca^{2+} channels, which have not yet been reported in guard cells, appear optimally suited for signal transduction in response to membrane depolarization in plant cells. Many physiological stimuli appear to trigger both Ca^{2+}-dependent and Ca^{2+}-independent transduction pathways as components of the specific signaling patterns. Further research concerning the regulation and cell-specific distribution of the ion channels described here, and other yet-to-be-identified ion channels, will enhance our understanding of their roles in signaling and provide probes for unraveling the intermediate signaling mechanisms used by plants in dynamic responses to the environment during growth and development.

Acknowledgements
This chapter is an updated version of an article that was originally published in *The Plant Cell*, July 1995 (Vol. 7, pp 833–844), copyright American Society of Plant Physiologists and is published here with the permission of both publishers and at the request of the editor of this volume. We thank Walter Kelly, Walter Gassmann and Rebecca Chasan for critical reading of this manuscript. Research in the laboratory of the authors was supported by grants from the USDA (Grant No. 92-37304-7757), DOE (DE-FG03-94-ER20148), National Science Foundation grant (MCB-9506191), and by an NSF Presidential Young Investigator Award to J.I. Schroeder, and an NSF Plant Biology postdoctoral fellowship to J.M. Ward.

References

Allan, A.C., Fricker, M.D., Ward, J.L., Beale, M.H. and Trewavas, A.J. (1994) Two transduction pathways mediate rapid effects of abscisic acid in *Commelina* guard cells. *Plant Cell* 6: 1319–1328.

Allen, G.J. and Sanders, D. (1994) Two voltage-gated, calcium release channels coreside in the vacuolar membrane of broad bean guard cells. *Plant Cell* 6: 685–694.

Allen, G.J. and Sanders, D. (1995) Calcineurin, a Type 2B protein phosphatase, modulates the Ca^{2+}-permeable slow vacuolar ion channel of stomatal guard cells. *Plant Cell* 7: 1473–1483.

Allen, G.J., Muir, S.R. and Sanders, D. (1995) Release of Ca^{2+} from individual plant vacuoles by both InsP3 and cyclic ADP-ribose. *Science* 268: 735–737.

Amodeo, G., Escobar, A. and Zieger, E. (1994) A cationic channel in the guard cell tonoplast of *Allium cepa*. *Plant Physiol.* 105: 999–1006.

Anderson, B.E., Ward, J.M. and Schroeder, J.I. (1994) Evidence for an extracellular reception site for abscisic acid in *Commelina* guard cells. *Plant Physiol.* 104: 1177–1183.

Armstrong, F., Leung, J., Grabov, A., Brearley, J., Giraudat, J. and Blatt, M.R. (1995) Sensitivity to abscisic acid of guard-cell K^+ channels is suppressed by abi1-1, a mutant *Arabidopsis* gene encoding a putative protein phosphatase. *Proc. Natl. Acad. Sci. USA* 92: 9520–9524.

Assmann, S.M. (1993) Signal transduction in guard cells. *Annu. Rev. Cell Biol.* 9: 345–375.

Bates, G.W. and Goldsmith, M.H.M. (1983) Rapid response of the plasma-membrane potential in oat coleoptiles to auxin and other weak acids. *Planta* 159: 231–237.

Bethke, P.C. and Jones, R.L. (1994) Ca^{2+}-calmodulin modulates ion channel activity in storage protein vacuoles of barley aleurone cells. *Plant Cell* 6: 277–285.

Blatt, M.R. (1990) Potassium channel currents in intact stomatal guard cells: Rapid enhancement by abscisic acid. *Planta* 180: 445–455.

Blatt, M.R. and Armstrong, F. (1993) K^+ channels of stomatal guard cells: abscisic acid-evoked control of the outward rectifier mediated by cytoplasmic pH. *Planta* 191: 330–341.

Blatt, M.R., Thiel, G. and Trentham, D.R. (1990) Reversible inactivation of K^+ channels of *Vicia* stomatal guard cells following photolysis of caged 1,4,5-trisphosphate. *Nature* 346: 766–769.

Bowler, C., Neuhaus, G., Yamagata, H. and Chua, N.-H. (1994) Cyclic GMP and calcium mediate phytochrome phototransduction. *Cell* 77: 73–81.

Cho, M.H. and Spalding, E.P. (1996) An anion channel in *Arabidopsis* hypocotyls activated by blue light. *Proc. Natl. Acad. Scj. USA* 93: 8134–8138.

Cole, K.S. and Curtis, H.J. (1938) Electrical impedance of *Nitella* during activity. *J. Gen. Physiol.* 22: 37–64.

Colombo, R., Cerana, R., Lado, P. and Peres, A. (1988) Voltage-dependent channels permeable to K^+ and Na^+ in the membrane of *Acer pseudoplatanus* vacuoles. *J. Membrane Biol.* 103: 227–236.

Cosgrove, D.J. and Hedrich, R. (1991) Stretch-activated chloride, potassium, and calcium channels coexisting in plasma membranes of guard cells of *Vicia faba* L. *Planta* 186: 143–153.

Coté, G.C. and Crain, R.C. (1994) Why do plants have phosphoinositides? *Bioessays* 16: 39–46.

Davies, E. (1987) Plant responses to wounding. *In*: P.K. Stumpf and E.E. Conn (eds): *The Biochemistry of Plants*, Vol. 12. Academic Press Inc., New York, pp 243–264.

DeSilva, D.L.R., Cox, R.C., Hetherington, A.M. and Mansfield, T.A. (1985) Synergism between calcium ions and abscisic acid in preventing stomatal opening. *New Phytol.* 101: 555–563.

Ebel, J., Bhagwat, A.A., Cosio, E.G., Feger, M., Kissel, U., Mithöfer, A. and Waldmüller, T. (1995) Elicitor-binding proteins and signal transduction in the activation of a phytoalexin defense response. *Can. J. Bot.* 73: 506–510.

Ehrhardt, D.W., Atkinson, E.M. and Long, S.R. (1992) Depolarization of alfalfa root hair membrane potential by *Rhizobium meliloti* Nod factors. *Science* 256: 998–1000.

Elzenga, J.T.M., Prins, H.B.A. and Van Volkenburgh, E. (1995) Light-induced membrane potential changes of epidermal and mesophyll cells in growing leaves of *Pisum sativum*. *Planta* 197: 127–134.

Esser, J.E., Liao, Y.-J. and Schroeder, J.I. (1996) Characterization of ion channel modulator effects on ABA- and malate-induced stomatal movements: strong regulation by kinase and phosphatase inhibitors, and relative insensitivity to mastoparans. *J. Exp. Bot.*; *in press*.

Fairley-Grenot, K.A. and Assmann, S.M. (1991) Evidence for G-protein regulation of inward K^+ current in guard cells of fava bean. *Plant Cell* 3: 1037–1044.

Finkelstein R.R. and Somerville C.R. (1990) Three classes of abscisic acid (ABA)-insensitive mutants of *Arabidopsis* define genes that control overlapping subsets of ABA responses. *Plant Physiol.* 94: 1172–1179.

Gaffey, C.T. and Mullins, L.J. (1958) Ion fluxes during the action potential in *Chara*. *J. Physiol.* 144: 505–524.

Gelli, A. and Blumwald, E. (1993) Calcium retrieval from vacuolar pools: Characterization of a vacuolar calcium channel. *Plant Physiol.* 102: 1139–1146.

Gilroy, S. and Jones, R.L. (1994) Perception of gibberellin and abscisic acid at the external face of the plasma membrane of barley (*Hordeum vulgare*) aleurone protoplasts. *Plant Physiol.* 104: 1185–1192.

Gilroy, S., Read, N.D. and Trewavas, A.J. (1990) Elevation of cytoplasmic calcium by caged calcium or caged inositol trisphosphate initiates stomatal closure. *Nature*, 346: 769–771.

Gilroy, S., Fricker, M.D., Read, N.D. and Trewavas, A.J. (1991) Role of calcium in signal transduction of *Commelina* guard cells. *Plant Cell* 3: 333–344.

Hedrich, R. and Neher, E. (1987) Cytoplasmic calcium regulates voltage-dependent ion channels in plant vacuoles. *Nature* 329: 833–837.

Hedrich, R., Busch, H. and Raschke, K. (1990) Ca^{2+} and nucleotide dependent regulation of voltage dependent anion channels in the plasmamembrane of guard cells. *EMBO J.* 9: 3889–3892.

Hedrich, R., Marten, I., Lohse, G., Dietrich, P., Winter, H., Lohaus, G. and Heldt, H.-W. (1994) Malate-sensitive anion channels enable guard cells to sense changes in the ambient CO_2 concentration. *Plant J.* 6: 741–748.

Hepler, P.K. and Wayne, R.O. (1985) Calcium and plant development. *Annu. Rev. Plant Physiol.* 36: 397–439.

Hille, B. (1992) *Ionic Channels of Excitable Membranes*, 2nd ed. Sinauer Associates Inc., Sunderand, MA.

Huang, J.W., Grunes, D.L. and Kochian, L.V. (1994) Voltage-dependent Ca^{2+} influx into right-side-out plasma membrane vesicles from wheat roots: Characterization of a putative Ca^{2+} channel. *Proc. Natl. Acad. Sci. USA* 91: 3473–3477.

Humble, G.D. and Raschke, K. (1971) Stomatal opening quantitatively related to potassium transport: Evidence from electron probe analysis. *Plant Physiol.* 48: 447–453.

Irving, H.R., Gehring, C.A. and Parish, R.W. (1992) Changes in cytoplasmic pH and calcium of guard cells precede stomatal closure. *Proc. Natl. Acad. Sci. USA* 89: 1790–1794.

Ishikawa, H., Aizawa, H., Kishira, H., Ogawa, T. and Sakata, M. (1983) Light-induced changes of membrane potential in guard cells of *Vicia faba*. *Plant Cell Physiol.* 24: 769–772.

Johannes, E., Brosnan, J.M. and Sanders, D. (1992) Parallel pathways for intracellular Ca^{2+} release from the vacuole of higher Plants. *Plant J.* 2: 97–102.

Kasamo, K. (1981) Effect of abscisic acid on the K^+ efflux and membrane potential of *Nicotiana tabacum* leaf cells. *Plant Cell Physiol.* 22: 1257–1267.

Keller, B.U., Hedrich, R. and Raschke, K. (1989) Voltage-dependent anion channels in the plasma membrane of guard cells. *Nature* 341: 450–453.

Keller, B.U., Hollmann, M., Heinemann, S. and Konnerth, A. (1992) Calcium influx through subunits GLUR1 and GLUR3 of kainate AMPA receptor channels is regulated by cAMP dependent protein kinase. *EMBO J.* 11: 891–896.

Kelly, W.B., Esser, J.E. and Schroeder, J.I. (1995) Effects of cytosolic calcium and limited, possible dual, effects of G protein modulators on guard cell inward potassium channels. *Plant J.* 8: 479–489.

Kinoshita, T., Nishimura, M. and Shimazaki, K. (1995) Cytosolic concentration of Ca^{2+} regulates the plasma membrane H^+-ATPase in guard cells of fava bean. *Plant Cell* 7: 1333–1342.

Kolb, H.A., Köhler, K. and Martinoia, E. (1987) Single potassium channels in the membranes of isolated mesophyll barley vacuoles. *J. Memb. Biol.* 95: 163–169.

Kuchitsu, K., Kikuyama, M. and Shibuya, N. (1993) N-acetylchitooligosaccharides, biotic elicitor for phytoalexin production, induce transient membrane depolarization in suspension-cultured rice cells. *Protoplasma* 174: 79–81.

Lemtiri-Chlieh, F. and MacRobbie, E.A.C. (1994) Role of calcium in the modulation of *Vicia* guard cell potassium channels by abscisic acid: a patch-clamp study. *J. Memb. Biol.* 137: 99–107.

Leonard, R.T. and Hepler, P.K. (eds) (1990) *Calcium in Plant Growth and Development*. American Society of Plant Physiologists, Rockville, MD.

Leung, J., Bouvier-Durand, M., Morris, P.-C., Guerrier, D., Chefdor, F. and Giraudat, J. (1994) *Arabidopsis* ABA response gene *ABI1*: Features of a calcium-modulated protein phosphatase. *Science* 264: 1448–1452.

Linder, B. and Raschke, K. (1992) A slow anion channel in guard cells, activation at large hyperpolarization, may be principle for stomatal closing. *FEBS Lett.* 131: 27–30.

Lunevsky, V.Z., Zherelova, O.M., Vostrikov, I.Y. and Berestovsky, G.N. (1983) Excitation of Characeae cell membrane as a result of activation of calcium and chloride channels. *J. Memb. Biol.* 72: 43–58.

MacRobbie, E.A.C. (1981) Effects of ABA on "isolated" guard cells of *Commelina communis* L. *J. Exp. Bot.* 32: 563–572.

MacRobbie, E.A.C. (1992) Calcium and ABA-induced stomatal closure. *Phil. Trans. R. Soc. Lond. B.* 338: 5–18.

MacRobbie, E.A.C. (1995) ABA-induced ion efflux in stomatal guard cells: Multiple actions of ABA inside and outside the cell. *Plant J.* 7: 565–576.

Mansfield, T.A., Hetherington, A.M. and Atkinson, C.J. (1990) Some current aspects of stomatal physiology. *Annu. Rev. Plant Physiol.* 41: 55–75.

Marshall, J., Corzo, A., Leigh, R.A. and Sanders, D. (1994) Membrane potential-dependent calcium transport in right-side-out plasma membrane vesicles from *Zea mays* L. roots. *Plant J.* 5: 683–694.

Marten, I., Lohse, G. and Hedrich, R. (1991) Plant growth hormones control voltage-dependent activity of anion channels in plasma membrane of guard cells. *Nature* 353: 758–762.

Marten, I., Busch, H., Raschke, K. and Hedrich, R. (1993) Modulation and block of the plasma membrane anion channel of guard cells by stilbene derivatives. *Eur. Biophys. J.* 21: 403–408.

Mathieu, Y., Kurkjian, A., Xia, H., Guern, J., Spiro, M., O'Neill, M., Albersheim, P. and Darvill, A. (1991) Membrane responses induced by oligogalacturonides in suspension-cultured tobacco cells. *Plant J.* 1: 333–343.

Mayer, M.L., MacDermott, A.B., Westbrook, G.L., Smith, S.J. and Barker, J.F. (1987) Agonist and voltage-gated calcium entry in cultured mouse spinal neurons under voltage clamp measured using arsenazo III. *J. Neurosci.* 7: 3230–3244.

McAinsh, M.R., Brownlee, C. and Hetherington, A.M. (1990) Abscisic acid-induced elevation of guard cell cytosolic Ca^{2+} precedes stomatal closure. *Nature* 343: 186–188.

McAinsh, M.R., Brownlee, C. and Hetherington, A.M. (1992) Visualizing changes in cytosolic-free Ca^{2+} during the response of stomatal guard cells to abscisic acid. *Plant Cell* 4: 1113–1122.

McAinsh, M.R., Webb, A.R., Taylor, J.E. and Hetherington, A.M. (1995) Stimulus-induced oscillations in guard cell cytosolic free calcium. *Plant Cell* 7: 1207–1219.

Meyer, K., Leube, M.P. and Grill, E. (1994) A protein phosphatase 2C involved in ABA signal transduction in *Arabidopsis thaliana*. *Science* 264: 1452–1455.

Mimura, T. and Shimmen, T. (1994) Characterization of the Ca^{2+}-dependent Cl^- efflux in perfused *Chara* cells. *Plant Cell Physiol.* 35: 793–800.

Neuhaus, G., Bowler, C., Kern, R. and Chua, N.-H. (1993) Calcium/calmodulin-dependent and -independent phytochrome signal transduction pathways. *Cell* 73: 937–952.

Nürnberger, T., Nennstiel, D., Jabs, T., Sacks, W.R., Hahlbrock, K. and Scheel, D. (1994) High affinity binding of a fungal oligopeptide elicitor to parsley plasma membranes triggers multiple defense responses. *Cell* 78: 449–460.

Outlaw, W.H., Jr. (1983) Current concepts on the role of potassium in stomatal movements. *Physiol. Plant.* 59: 302–311.

Outlaw, W.H., Jr., Manchester, J. and Brown, P.H. (1981) High levels of malic enzyme activities in *Vicia faba* L. epidermal tissue. *Plant Physiol.* 66: 1047–1051.

Pantoja, O., Dainty, J. and Blumwald, E. (1992) Cytoplasmic chloride regulates cation channels in the vacuolar membrane of plant cells. *J. Memb. Biol.* 125: 219–229.

Piñeros, M. and Tester, M. (1995) Characterization of a voltage-dependent Ca^{2+}-selective channel from wheat roots. *Planta* 195: 478–488.

Racusen, R.H. and Satter, R.L. (1975) Rhythmic and phytochrome-regulated changes in transmembrane potential in *Samanea* pulvini. *Nature* 255: 408–410.

Ranjeva, R., Graziana, A., Mazars, C. and Thuleau, P. (1992) Putative L-type calcium channels in plants: Biochemical properties and subcellular localisation. *In*: D.T. Cooke and D.T. Clarkson (eds): *Transport and Receptor Proteins of Plant Membranes*. Plenum Press, New York, pp 145–153.

Raschke, K. (1979) Movements of stomata. *In*: H.A. Feinleib (ed.): *Encyclopedia of Plant Physiology*. Springer-Verlag, Berlin, pp 384–441.

Schmidt, C. and Schroeder, J.I. (1994) Anion selectivity of slow anion channels in *Vicia faba* guard cells: large nitrate permeability. *Plant Physiol.* 106: 383–391.

Schmidt, C. Schelle, I., Liao, Y.J. and Schroeder, J.I. (1995) strong regulation of slow anion channels and abscisic acid signaling in guard cells by phosphorylation and dephosphorylation events. *Proc. Natl. Acad. Sci. USA* 92: 9535–9539.

Schneggenburger, R., Zhou, Z., Konnerth, A. and Neher, E. (1993) Fractional contribution of calcium to the cation current through glutamate receptor channels. *Neuron* 11: 133–143.

Schroeder, J.I. (1988) K^+ transport properties of K^+ channels in the plasma membrane of *Vicia faba* guard cells. *J. Gen. Physiol.* 92: 667–683.

Schroeder, J.I. (1995) Anion channels as central mechanisms for signal transduction in guard cells and putative functions in roots for plant-soil interaction. *Plant Molec. Biol.* 28: 353–361.

Schroeder, J.I. and Hagiwara, S. (1989) Cytosolic calcium regulates ion channels in the plasma membrane of *Vicia faba* guard cells. *Nature* 338: 427–430.

Schroeder, J.I. and Hagiwara, S. (1990) Repetitive increases in cytosolic Ca^{2+} of guard cells by abscisic acid, activation of non-selective Ca^{2+} permeable channels. *Proc. Natl. Acad. Sci. USA* 87: 9305–9309.

Schroeder, J.I. and Hedrich, R. (1989) Involvement of ion channels and active transport in osmoregulation and signaling of higher plant cells. *Trends Biochem. Sci.* 14: 187–192.

Schroeder, J.I. and Keller, B.U. (1992) Two types of anion channel currents in guard cells with distinct voltage regulation. *Proc. Natl. Acad. Sci. USA* 89: 5025–5029.

Schroeder, J.I., Hedrich, R. and Fernandez, J.M. (1984) Potassium-selective single channels in guard cell protoplasts of *Vicia faba*. *Nature* 312: 361–362.

Schroeder, J.I., Raschke, K. and Neher, E. (1987) Voltage dependence of K^+ channels in guard cell protoplasts. *Proc. Natl. Acad. Sci. USA* 84: 4108–4112.

Schroeder, J.I., Schmidt, C. and Scheaffer, J. (1993) Identification of high-affinity slow anion channel blockers and evidence for stomatal regulation by slow anion channels in guard cells. *Plant Cell* 5: 1831–1841.

Schulz-Lessdorf, B., Dietrich, P., Marten, I., Lohse, G., Busch, H. and Hedrich, R. (1994) Coordination of plasma membrane and vacuolar membrane ion channels during stomatal movement. *In: Membrane Transport in Plants and Fungi*. SEB Symposium 48. The Company of Biologists Ltd., pp 99–112.

Schulz-Lessdorf, B. and Hedrich, R. (1995) Protons and calcium modulate SV-type channels in the vacuolar-lysosomal compartement – channel interaction with calmodulin inhibitors. *Planta* 197: 655–671.

Schwartz, A., Ilan, N., Grantz, D.A. (1988) Ca^{2+} modified stomatal responses to light, KCl and CO_2. *Plant Physiol.* 87: 583–587.

Schwartz, A., Wu, W.H., Tucker, E.B. and Assmann, S.M. (1994) Inhibition of inward K^+ channels and stomatal response by abscisic acid – an intracellular locus of phytohormone action. *Proc. Natl. Acad. Sci. USA* 91: 4019–4023.

Schwartz, A., Ilan, N., Schwarz, M., Scheaffer, J., Assmann, S.M, and Schroeder, J.I. (1995) Anion-channel blockers inhibit S-type anion channels and abscisic acid responses in guard cells. *Plant Physiol.* 109: 651–658.

Shacklock, P.S., Read, N.D. and Trewavas, A.J. (1992) Cytosolic free calcium mediates red light induced photomorphogenesis. *Nature* 358: 753–755.

Shiina, T. and Tazawa, M. (1988) Ca^{2+}-dependent Cl^- efflux in tonoplast-free cells of *Nitellopsis obtusa*. *J. Memb. Biol.* 106: 135–139.

Shimmen, T., Mimura, T., Kikuyama, M. and Tazawa, M. (1994) Characean cells as a tool for studying electrophysiological characteristics of plant cells. *Cell Struct. Func.* 19: 263–178.

Spalding, E.P. and Cosgrove, D.J. (1989) Large plasma-membrane depolarization precedes rapid blue-light-induced growth inhibition in cucumber. *Planta* 178: 407–410.

Sze, H., Ward, J.M. and Lai, S. (1992) Vacuolar H^+-translocating ATPases from plants: Structure, function, and isoforms. *J. Bioenerg. Biomemb.* 24: 371–381.

Tazawa, M. Shimmen, T. and Mimura, T. (1987) Membrane control in the characeae. *Annu. Rev. Plant Physiol.* 38: 95–117.

Thiel, G., MacRobbie, E.A.C. and Blatt, M.R. (1992) Membrane transport in stomatal guard cells: the importance of voltage control. *J. Membrane Biol.* 126: 1–18.

Thuleau, P., Ward, J.M., Ranjeva, R. and Schroeder, J.I. (1994) Voltage-dependent calcium-permeable channels in the plasma membrane of a higher plant cell. *EMBO J.* 13: 2970–2975.

Tyerman, S.D. (1992) Anion channels in plants. *Annu. Rev. Plant Physiol.* 43: 351–373.

Ullrich, C.I. and Novacky, A.J. (1991) Electrical membrane properties of leaves, roots, and single root cap cells of susceptible *Avena sativa* – effect of victorin-C. *Plant Physiol.* 95: 675–681.

Ward, J.M. and Schroeder, J.I. (1994) Calcium-activated K^+ channels and calcium-induced calcium release by slow vacuolar ion channels in guard cell vacuoles implicated in the control of stomatal closure. *Plant Cell* 6: 669–683.

Ward, J.M., Pei, Z.-M. and Schroeder, J.I. (1995) Roles of ion channels in initiation of signal transduction in higher plants. *Plant Cell* 7: 833–844.

Weiser, T., Blum, W. and Bentrup, F.-W. (1991) Calmodulin regulates the Ca^{2+}-dependent slow-vacuolar ion channel in the tonoplast of *Chenopodium rubrum* suspension cells. *Planta* 185: 440–442.

Wildon, D.C., Thain, J.F., Minchin, P.E.H., Gubb, I.R., Reilly, A.J., Skipper, Y.D., Doherty, H.M., O'Donnell, P.J. and Bowles, D.J. (1992) Electrical signaling and systemic proteinase inhibitor induction in the wounded plant. *Nature* 360: 62–65.

Williamson, R.E. and Ashley, C.C. (1982) Free Ca^{2+} and cytoplasmic streaming in the alga *Chara*. *Nature* 296: 647–651.

Zimmermann, S., Thomine, S., Guern, J. and Barbier-Brygoo, H. (1994) An anion current at the plasma membrane of tobacco protoplasts shows ATP-dependent voltage regulation and is modulated by auxin. *Plant J.* 6: 707–716.

Signal Transduction in Plants
P. Aducci (ed.)
© 1997 Birkhäuser Verlag Basel/Switzerland

ABA signaling in plant development and growth

T.L. Thomas, H.-J. Chung and A.N. Nunberg

Department of Biology, Texas A&M University, College Station, TX 77843-3258, USA

Introduction

Abscisic acid (ABA) is implicated in a number of developmental events during the life cycle of higher plants. Not only is ABA involved in mediating the response to a number of environmental stresses, including drought, cold and salt, it also plays a significant role in embryo development and seed maturation. These specific, pleiotropic responses are often reflected in differential patterns of gene expression. There is comparatively little information on the mechanism(s) of ABA perception and the subsequent downstream signaling pathway(s) involved that lead to a physiological or genetic response. However, recent progress in identifying and utilizing mutations that affect ABA synthesis and perception coupled with the analysis of ABA induced stomatal guard cell closure and the analysis of genes and their promoters that respond to ABA is leading to new insights into the ABA signaling pathway(s). Below, we present an eclectic review of results in these three areas and end with a prospectus on future directions.

Molecular genetics of ABA mediated signal transduction

Biochemical characterization of ABA signaling components have been largely unsuccessful. A more useful approach has been the identification and characterization of mutants defective in the synthesis and perception of ABA. ABA mutants have been identified in a number of plant species including *Arabidopsis*, maize, and tomato (reviewed in McCarty, 1995; Giraudat et al., 1994). Recent and extensive studies have focused on the ABA deficient and insensitive loci of *Arabidopsis* and the maize ABA insensitive mutant, *vp1*.

Two classes of Arabidopsis *ABA mutants:* aba *and* abi

aba (ABA-deficient) mutants of *Arabidopsis* have low levels of ABA due to attenuated rates of accumulation in both seeds and leaves. *abi* (ABA-insensitive) mutants are impaired or deficient in various responses regulated by ABA such as failure to achieve dormancy and reduced stomatal closure. Initially, ABA insensitive mutants of *Arabidopsis* were identified from a population of mutagenized seeds based on their ability to germinate and grow in the presence of ABA (Koornneef et al., 1982; 1984). Three loci were isolated, *abi1*, *abi2*, and *abi3*. Under water stress, *abi1* and *abi2* produced a wilty phenotype, while *abi3* plants appeared normal. All three mutants result in non-dormant seeds. In severe alleles of *abi3*, seed development is normal up to the matu-ration stage, at which point, seeds accumulate low levels of storage proteins and lipids, do not degrade accumulated chlorophyll from earlier development, and become desiccation intolerant (Nambara et al., 1994; Ooms et al., 1993; Nambara et al., 1992). These seeds may be rescued by imbibition before desiccation but are usually non-dormant, whereas wild type *Arabidopsis* seeds become dormant upon maturation. While all three *abi* loci are involved in overlapping pathways, *abi3* apparently is seed-specific (Finkelstein and Somerville, 1990). The *ABI3* locus has been cloned; it encodes a novel protein with regions similar to the maize *VP1* gene, a transcriptional activator/repressor (see below).

Recent studies have focused on the role of *ABI3* during seed development, particularly on the effect of *abi3* on gene expression. For example, Parcy et al. (1994) investigated the regulation of 18 genes in developing *Arabidopsis* wild-type and *abi3* seeds. These genes were divided into four classes based on their temporal expression: Class 1 (early genes), *MAT* (maturation) genes, *LEA* (late embryogenesis abundant) and *LEA-A* genes. In the *abi3* background, expression of the Class 1 genes during early embryogenesis was not altered, but three of the Class 1 genes showed expression extending further into development. These results suggest that *ABI3* may act to repress genes during late embryo development. Several *MAT* class genes were repressed in an *abi3* background (Nambara et al., 1992; Finkelstein and Somerville, 1990). The *MAT* genes in Parcy et al. (1994) included napin 3 (a 2S albumin), cruciferin c (a 12S globulin), vicilin, and oleosin. Expression of the 2S albumin and cruciferin c was severely repressed, while vicilin and oleosin expression remained near wild-type levels. Both *LEA* and *LEA-A* genes were repressed in the *abi3* background. It is noteworthy that not all seed storage protein genes are responsive to ABA, and regulation *via* an *ABI3* dependent pathway may correlate with ABA inducible expres-sion. Furthermore, an underlying assumption is that ABA regulated expression of seed storage proteins is due to the perception of ABA within the developing embryo (Thomas, 1993). In an ABA deficient background (*aba1*), napin 3 and cruciferin c expression is normal; while two *LEA* genes were repressed (Parcy et al., 1994). It would appear that these seed storage protein genes are not regulated by ABA content but by the presence of an ABA signaling component. However, since the levels of ABA are not reduced to zero in the ABA deficient background, it is still possible that ABA content can induce napin 3 and cruciferin c transcripts but is insufficient for the induction of the ABA responsive *LEA* genes.

ABI3 apparently functions not only in ABA signaling, but also in the expression of several genes in the absence of ABA. Because of sequence similarity with *VP1*, it is hypothesized that ABI3 is a transcription factor; however, neither has been shown to interact directly with DNA. ABI3 is expressed exclusively in the seed and is capable of inducing napin-3 and cruciferin-c expression in germinating seedlings exposed to ABA when ectopically expressed (Parcy et al., 1994). It is possible that ABI3, as well as VP1, may be part of a higher order regulatory network that intersects the ABA/seed regulatory network.

Severe alleles of *abi3* illustrate the pivotal role it plays in proper embryo development. Parcy et al. (1994) showed that ABI3 is a positive regulator of several genes involved in maturation as well as possibly being a repressor of genes expressed early in embryogenesis. ABI3 also represses germination in developing embryos. Developing seeds carrying a severe allele of *abi3* exhibit characteristics of germinating seedlings. Nambara et al. (1995) showed that *abi3* embryos have meristems that resemble wild type germinating seedlings and development of vascular tissue has progressed to a more mature state. The chlorophyll a/b binding protein is expressed at the same levels as 3-day imbibed wild-type seeds. The initiation of a germination pathway during embryo development in *abi3* embryos is similar to that observed for *leafy cotyledon* (*lec1*) and *fusca3* (*fus3*) mutant embryos (Keith et al., 1994; Meinke et al., 1994). Unlike *abi3* embryos, *lec1* and *fus3* embryos respond to ABA in culture and also have altered cotyledon morphology; furthermore, studies of double mutants have shown that *abi3*, *leafy cotyledon* and *fusca* mutants act in different pathways (Bäumlein et al., 1994; Meinke et al., 1994).

Analysis of plants defective in ABA perception or synthesis is an effective approach to elucidate the role of ABA mediated signal transduction in cold or drought acclimation (Mäntylä et al., 1995; Lång et al., 1994; Nordin et al., 1993; Lång and Palva, 1992; Meurs et al., 1992). For example, *aba-1* and *abi1* were severely impaired in their ability to cold-acclimate (Mäntylä et al., 1995). The impaired cold-acclimation phenotype of *aba-1* could be rescued with exogenous ABA (Heino et al., 1990); this was not the case with *abi1* (Mäntylä et al., 1995). Similar results were observed in drought-induced gene(s)/clone(s). mRNA accumulation of two drought-induced genes, *AtDi8* and *AtDi21* was abolished in *aba1* and *abi1-1*, but not in the *abi3-4* mutant (Gosti et al., 1995).

In our laboratory, we have investigated the impact of ABA mutants *abi1-1* and *abi3-1* on the expression of the carrot *LEA*-class gene *Dc3* (Thomas, 1993; Vivekananda et al., 1992; Seffens et al., 1990). We crossed transgenic *Arabidopsis* containing the *Dc3* full-length promoter to *abi1-1* and *abi3-1* mutants to determine their effect on *Dc3* expression during embryogenesis or under conditions of drought or exposure to exogenous ABA. Both mutants are impaired in ABA inducibility, but *abi1-1* has been implicated in ABA induction in seed and vegetative tissues while *abi3-1* is apparently only involved in seed development. *Dc3* expression in mature seeds and vegetative tissues exposed to exogenous ABA or drought was observed by determining the expression of the GUS reporter gene driven by the *Dc3* promoter in *abi1* and *abi3* mutant backgrounds. *Dc3*-driven GUS expression during seed development was significantly reduced in both mutants; whereas ABA induction in vegetative tissues was not affected. Previously, we reported that drought stress induced *Dc3*-driven GUS expression in transgenic tobacco seedlings as well (Vivekananda et al., 1992). Surprisingly, the *Dc3*/GUS *Arabidopsis* plants did not show any drought effect in wild type or *abi* plants; rather expression of *Dc3*/GUS was reduced in both wild type and mutants. This suggests that *Arabidopsis* plants may not contain analogous factors to those in tobacco to recognize *cis*-elements in the *Dc3* promoter required for drought response. Our results suggest that there are at least two signal transduction pathways involved in *Dc3* expression. One is an ABA-dependent pathway that is involved in *Dc3* expression during embryogenesis. Expression of *Dc3* in vegetative tissues by exogenous ABA apparently requires another signal transduction pathway in which ABI1 and ABI3 are not required directly.

The maize viviparous (vp) *mutants*

The ABA insensitive mutant of maize, *viviparous 1* (*vp1*), shares many similarities to the *abi-3* mutant of *Arabidopsis*. Both produce non-dormant seeds that do not undergo normal maturation, and both phenotypes are restricted to the seed. Unlike *Arabidopsis*, *vp1* embryos proceed directly to germination (vivipary). Several seed storage protein genes, as well as other maturation genes are down regulated in *vp1* embryos (Paiva and Kriz, 1994). As with the study of Parcy et al. (1994), the major seed storage proteins of maize are not significantly reduced in ABA deficient maize kernels, but ABA does accumulate in these kernels, albeit at greatly reduced levels (Paiva and Kriz, 1994). In both cases, it may be that reduced ABA content, although insufficient to prevent precocious germination and dormancy, is sufficient to induce seed storage gene expression.

The *VP1* locus has been cloned and shown to encode a transcriptional activator (Hattori et al., 1992; McCarty et al., 1991). Like ABI-3, VP1 can also act to repress at least one aspect of the germination pathway, and apparently this repression does not require ABA (Hoecker et al., 1995).

Similar gene products have been cloned from rice, tobacco and bean (Bobb et al., 1995; Hattori et al., 1994; Phillips and Conrad, 1994). It is possible that the seed-specific ABA signaling pathway is conserved among higher plants and that elements of this pathway have been recruited for ABA-independent gene regulation that coordinates embryo maturation, seed dormancy, and germination. Alternatively and perhaps more likely, the seed/embryo program evolved first and the ABA components were recruited by some, but not all genes. However, ABA signaling in seeds is complex. Isolation of two new *Arabidopsis* ABA insensitive mutants demonstrates multiple pathways and at least one of these pathways does not overlap ABI-3 mediated ABA signaling (Finkelstein, 1994).

ABA induced stomatal guard cell closure

Plants balance water loss with gas exchange through openings in the adaxial surface of the leaves called stomata. Stomatal openings allow atmospheric CO_2 to enter leaves for carbon fixation; these openings also allow O_2 to escape. Water collaterally evaporates through open stomata. Two specialized cells (guard cells) form the stomatal opening. Guard cells regulate the aperture of the stomata by changes in cell turgor. When cell turgor is reduced the aperture is closed; when turgor is restored the stoma is open. Guard cells respond to a number of signals to regulate stomatal aperture; one of these signals is ABA. Guard cells are a useful model in the analysis of ABA action because the closure response can be detected within minutes of treatment. ABA induces the closing of stomatal guard cells, but it also inhibits opening.

The physiological events that are involved in stomatal guard cell closing will be briefly reviewed followed by more recent advances in some key areas of the signal cascade. (For detailed reviews of the electrophysiological changes that occur during ABA induced stomatal guard cell closure see Giraudat et al., 1995, 1994; Blatt and Thiel, 1993.)

Closure of stomatal guard cells mediated by ABA involves an efflux of K^+ and anions from the tonoplast and cytoplasm across the plasma membrane which results in loss of turgor. A sustained electrochemical gradient is needed for this efflux to occur. When guard cells are exposed to ABA, the first electrical change is an initial depolarization of the plasma membrane which is a result of a net influx of positive charge (Thiel et al., 1992). During this same time period, there is an increase in cytosolic Ca^{2+} through the activation of nonselective Ca^{2+} channels (Schroeder and Hagiwara, 1990). Changes in pH also occur upon exposure to ABA, leading to alkalinization of the cytosol (Irving et al., 1992). Voltage and Ca^{2+} gated anion channels are activated that produce the electrochemical force that drives K^+ through the outward rectifying K^+ channels.

ABA perception

One of the most basic questions concerning ABA signaling in guard cells is still unanswered. Where is the receptor(s) for ABA located? Hornberg and Weiler (1984) reported ABA binding activity in the plasma membrane of *Vicia faba* guard cells and that this binding site faced outward. Although this result was never duplicated, it gave the first indication that the ABA receptor may reside on the plasma membrane. Corroborating evidence has only recently become available to support the hypothesis of a plasma membrane ABA receptor. To address the question of a cytosolic receptor, Anderson et al. (1994) microinjected ABA into the cytoplasm of *Commelina* guard cells and determined that at an extracellular pH of 6.15 or 8.0; microinjected ABA did not inhibit stomatal opening. Under the same conditions, externally applied ABA did inhibit stomatal opening. Interestingly, externally applied ABA was more effective at inhibiting the stomatal guard cells at pH 6.15 where a large majority of the molecule is protonated and easily diffusible through the plasma membrane (Kaiser and Hartung, 1981). Thus, it is possible that there is also a cytosolic ABA receptor because the injected ABA in the study was buffered to pH 7.2 so that changes in the cellular pH may have masked some effects.

Barley aleurone protoplasts are useful in studying gibberellic acid (GA) induced gene expression as well as the antagonistic interaction between GA and ABA. Gilroy and Jones (1994) asked where ABA and gibberellic acid perception occurred in barley aleurone protoplasts. Protoplasts were immobilized in an agarose/starch matrix which allowed manipulation of individual cells. An α-amylase promoter fused to β-glucuronidase (GUS) was used as a reporter for GA induced gene expression. α-amylase secretion was monitored by starch-iodine staining. GA induced changes in GUS expression as well as alterations in starch metabolism were observed when the hormone was applied externally, but injection of up to 250 μM GA had no apparent effect. Similarly, ABA antagonism was observed only when ABA was applied externally.

Both studies strongly point toward a plasma membrane ABA receptor. However, there are indications that cytosolic ABA also plays a role in guard cell signaling. For example, the efficiency of ABA inhibition of stomatal opening is pH dependent (Anderson et al., 1994). Also, the release of caged-ABA by photolysis and the microinjection of ABA can lead to stomatal closure (Allan et al., 1994; Schwartz et al., 1994). Furthermore, radioactive tracer studies suggest that the rate of ABA induced $^{86}Rb^+$ efflux is influenced by both extracellular and intracellular ABA (MacRobbie, 1995). Exactly where the perception of ABA occurs in guard cells is still unclear and may not be resolved until an ABA receptor(s) has been characterized at the molecular and cellular level.

The role of Ca^{2+} in ABA signaling

One of the first detectable events in ABA signaling in guard cells is an increase in cytoplasmic Ca^{2+}. There are a number of ways in which Ca^{2+} has been implicated in ABA signaling. First, the depolarization that is initially detected in ABA signaling can be due, in part, to the uptake of Ca^{2+} through non-selective channels. Release of caged Ca^{2+} as well as caged inositol 1,4,5-triphosphate (IP$_3$) is sufficient for stomatal closure (Gilroy et al., 1990). This also suggests that the release of Ca^{2+} from internal stores may play a role. This observation is interesting since phosphoinositides, phosphorylated at the 3 and 4 positions, and inositol phosphates were identified in stomatal guard cells of *Commelina* (Parmar and Brearley, 1993). This hypothetical pathway may be related to G protein mediated signal transduction in animals. Release of Ca^{2+} into the cytosol can also induce voltage-gated anion channels which is the apparent driving force for K$^+$ efflux (Schroeder and Hagiwara, 1989).

More recently, another role for Ca^{2+} has been suggested. Kinoshita et al. (1995) described reversible inhibition of the plasma membrane H$^+$-ATPase by cytsolic Ca^{2+} in *Vicia faba* guard cells at physiological levels. It was already established that the plasma membrane H$^+$-ATPase drives K$^+$ uptake in guard cells for stomatal opening (Briskin and Hanson, 1992; Hedrich and Schroeder, 1989). A role for the plasma membrane H$^+$-ATPase in stomatal closure had been lacking even though inhibition results in a depolarized membrane potential (Tazawa et al., 1987). Plasma membrane H$^+$-ATPase maintains a polarized membrane potential. In order to initiate the cascade of events that lead to stomatal closure, membrane potential must be depolarized (see below). It is currently unclear whether the initial influx of Ca^{2+} is sufficient to depolarize the membrane. Inhibition of the H$^+$-ATPase may contribute significantly to the initial membrane depolarization. Initial depolarization of the plasma membrane associated with the increase in Ca^{2+} (net increase in positive charge) may therefore be attributed, in part, to the inhibition the H$^+$-ATPase by Ca^{2+}.

Allan et al. (1994) noted the variation of published data concerning ABA induced increases in Ca^{2+} in guard cells as well as the fact that, in some cases, there is no detectable Ca^{2+} increase even though closure was observed. This raises questions about the relationship between ABA and Ca^{2+}. However, Allan and coworkers demonstrated that increases in Ca^{2+} due to the release of caged ABA correlated with the temperature regime plants received. At lower temperatures, no detectable increase in Ca^{2+} occurred, even though the stomata closed at rates comparable to higher temperatures. This suggests that ABA induces stomatal closure through a Ca^{2+} dependent as well as a Ca^{2+} independent pathway. It remains to be seen at what point, if ever, the two pathways converge.

Ion channels

Regulation of stomatal aperture occurs by changes in guard cell turgor which are brought about by changes in solute concentration. This is due in part to the efflux or uptake of K^+. Stomatal closure is the result of reduced turgor in the guard cells which is accompanied by a large efflux of K^+ as well as anions. During stomatal closure, the plasma membrane is depolarized for a relatively long period of time, allowing a large efflux of K^+ through outward rectifying K^+ channels as well as anions through their appropriate channels. Patch-clamp studies have identified anion currents in the plasma membrane of guard cells (Schroeder and Hagiwara, 1989). These anion channels can be divided into two classes, slow (S) and rapid (R). S-type channels are voltage regulated and allow large and sustained anion efflux (Schroeder and Keller, 1992). They are therefore implicated in providing the long-term membrane depolarization which drives K^+ efflux. Anion channel blockers, specific for S-type channels, applied extracellularly can block the closure of stomata in the presence of ABA and malate as well as promoting stomatal opening (Schroeder et al., 1993). These channel blockers also can inhibit ABA induced stomatal closure in the absence of malate and reverse ABA inhibition of stomatal opening (Schwartz et al., 1995). This suggests that stomatal opening involves the inhibition of anion channels in the plasma membrane which would ordinarily allow anions to migrate out and inhibit ion uptake.

S-type anion channels are emerging as important components for stomatal aperture control, but the regulation of these anion channels is poorly understood. Rigorous control of anion channels may be mediated by phosphorylation events. S-type anion currents can be greatly (90%) inhibited by either removal of cytosolic ATP or by kinase inhibitors (Schmidt et al., 1995). These currents can also be enhanced by the application of okadaic acid, a phosphatase inhibitor (Schmidt et al., 1995). Thus, there appears to be both a kinase and phosphatase activity in guard cells that regulates S-type anion channels in the plasma membrane. This level of regulation may account for both Ca^{2+}-dependent and independent pathways of ABA induced stomatal closure. The target of the kinase/phosphatase has not been identified; furthermore, it remains to be seen if the channel itself is modified in this way.

Interestingly, K^+ outward rectifying channels are also regulated by protein phosphorylation. Armstrong et al. (1995) transferred the dominant ABA insensitive mutant *abi1* (Koorneef et al., 1982) of *Arabidopsis* into tobacco to study the effects this mutation had on stomatal control. A 6.5 kb genomic fragment containing the *abi1-1* allele was transferred to tobacco *via Agrobacterium* transformation. Transgenic tobacco containing the mutant gene display the same wilty phenotype. Electrophysiological studies demonstrated that outward rectifying K^+ currents are reduced and are insensitive to exogenous applied ABA even though the transgenic lines have normal response to ABA in terms of anion current as well as alkalinization of the cytosol (Armstrong et al., 1995). Since the *ABI1* gene encodes a putative protein phosphatase (Leung et al., 1994), the role of protein phosphorylation in regulating K^+ currents was examined. Normal sensitivity to ABA, in terms of K^+ current and stomatal closure, was restored when broad ranged antagonists of protein kinases were used.

The above observations link protein phosphorylation with the regulation of ion currents in guard cells, and there could be at least two kinase/phosphatase pathways involved in regulating stomatal aperture. This is important since it may provide a link between ABA signaling in physiological events (like stomatal closing) and changes in gene expression. ABA is known to affect protein phosphorylation (Koontz and Choi, 1993; Anderberg and Walker-Simmons, 1992; Chapman et al., 1975). ABI1 is a putative protein phosphatase and is involved in regulating stomatal aperture. It is also known to be involved in the regulation of a number of ABA-responsive genes (see below). In both cases, the effects of ABI1 are limited. Not all currents in guard cells are affected, and likewise not all ABA-responsive genes are affected by ABI1.

Guard cells are an excellent model to study common components of ABA signaling because of their dramatic physiological responses. Furthermore, they are capable of responding to ABA by inducing expression of specific genes (Taylor et al., 1995). Future studies utilizing this powerful tool should elucidate ABA mediated signal transduction involved in the physiological responses and changes in gene expression in individual cells with different genetic backgrounds.

ABA-responsive genes and their promoters

Plants respond physiologically and biochemically to adverse environmental conditions such as drought, cold, or high salinity, and developmental cues such as desiccation and dormancy; these specific and often pleiotropic responses are reflected in differential patterns of gene expression. An increase in endogenous ABA is frequently associated with drought-, cold- and salt-stress or during seed desiccation; these responses to environmental or developmental cues may share some signal transduction components in addition to ABA. This hypothesis is supported by extensive studies of cold-, drought-, or ABA-regulated genes of a number of plant species (Tab. 1).

Diverse genes respond to developmental and environmental cues and to ABA

During seed development, ABA regulates several physiological processes such as storage protein synthesis (Naito et al., 1994; Paiva and Kritz, 1994; Plant et al., 1994), inhibition of precocious

Table 1. Diversity of putative ABA responsive elements

Gene	Species	Induction of endogenous gene[a]		Em-like element[b]	ABA Response[c]	Comments[d]	Reference
Em	Wheat	ABA	salt	GACACGTGGCG CACACGTGCCG	+ +	5' untranslated region increases the ABA response	Marcotte et al. (1989)
HVA1	Barley	ABA salt	LT heat	CGTACGTGCA CCTACGTGGCG GCAACGTGTC	− + +	Work together and may require a coupling element	Straub et al. (1994)
HVA22	Barley	ABA		CGCACGTGTC GCCACGTACA	+ +	Requires coupling element CE1, TGCCACCGG	Shen and Ho (1995)
rab16A	Rice	ABA	salt	CGTACGTGGC CGCCGCGCC/T	+ +		Mundy et al. (1990)
Osem	Rice	ABA		TACGTGTC GACGTGTC GACACGTA	+ + +	Contains additional ABA response motif, CGGCGGCCTCGCCACG	Hattori et al. (1995)
Dc3	Carrot	ABA	DR	TTTCGTGT AACGTGTT GACGTGTA TACGTGTC	+ − − −	Requires proximal promoter region	Chung et al. (in preparation)
HaG3	Sunflower	ABA	DR	TACGAAC	+	Interacts with proximal promoter region	Nunberg et al. (1994)
rab17	Maize	ABA	DR	ACGTGGC	+		Vilardell et al. (1991)
rab28	Maize	ABA	DR	CCACGTGC	+		Pla et al. (1993)
CdeT27-45	C. plantagineum	ABA		TACACGTGTGC AACACGTACGA GGCACGTATGT TACACGTTTCA	− − − +	Contains ABA-inducible binding motif, AGCCCA	Michel et al. (1993) Neslon et al. (1994)
CdeT6-19	C. plantagineum	ABA	DR	ATACGTGT ACACGTAA GCACGTGG AAACGTGT	+ + − −	Responds in tobacco protoplasts	Michel et al. (1994)

Table 1. (continued)

Gene	Species	Induction of endogenous gene[a]		Em-like element[b]	ABA Response[c]	Comments[d]	Reference
rd22	*Arabidopsis*	ABA	DR	GCACGTAG	+	Contains MYB, MYC binding sites	Iwasaki et al. (1995)
rd29A (lti78)	*Arabidopsis*	ABA DR	LT salt	ATACGTGTC	+	Contains conserved DRE, TACCGACAT	Yamaguchi-Shinozaki and Shinozaki (1994)
rd29B (Lti65)	*Arabidopsis*	ABA	DR	GTACGTGTC AGACGTGGC		Identified by sequence comparison	Yamaguchi-Shinozaki and Shinozaki (1994)
kin1	*Arabidopsis*	ABA DR	LT	ACACGTGA CCACGTGG ACTCGTGG		Identified by sequence comparison, contains conserved DRE	Wang et al. (1995)
cor6.6	*Arabidopsis*	ABA DR	LT	ACACGTGA CCACGTGG ACACGTGG		Identified by sequence comparison, contains conserved DRE	Wang et al. (1995)
cor15a	*Arabidopsis*	ABA DR	LT	ATACGTGT ACACGTGG ACACGTGA	− + +		Baker et al. (1994)
rab18	*Arabidopsis*	ABA DR	LT	GTACGTGT TTACGTGT		Identified by sequence comparison	Lång and Palva (1992)

[a] ABA, abscisic acid; DR, drought; LT, low temperature
[b] Some similarity with G-box or ACGT core sequence
[c] ABA response of element in transgenic plants or in protoplasts. No entry indicates "not determined" or "results unknown"
[d] DRE, drought response element

germination, dormancy onset and maintenance, and desiccation tolerance acquisition (Black, 1991). ABA levels peak in the developing seed just prior to desiccation and correlate with synthesis of a set of proteins encoded by a phylogenetically conserved group of genes known as late embryogenesis abundant (*LEA*) genes (Dure et al., 1989). Some, but not all, *LEA* genes identified so far can be induced in young, excised embryos by exogenous ABA (Dure, 1993) suggesting that ABA may affect seed development and gene expression differentially during this late phase of embryogenesis. Many ABA-regulated genes expressed in seeds, including *LEA* genes, have been cloned from cotton, rape, soybean, sunflower, rice, barley, maize, wheat, and *Arabidopsis* (reviewed in Quatrano et al., 1993; Skriver and Mundy, 1990).

Drought stress elicits many physiological changes such as stomatal closure (Zeevaart et al., 1991), reduced hypocotyl elongation (Creelman et al., 1990), as well as significant biochemical changes, e.g., accumulation of newly synthesized polypeptides (Reviron et al., 1992; Bray, 1991), including cell-wall proteins (Covarrubias et al., 1995; Wu et al., 1995), and soluble carbohydrates (Creelman, 1990). Application of exogenous ABA can also elicit similar drought effects on plants, implying that ABA is involved in drought acclimation as well. Drought-induced genes have been identified in several species and have been classified into four different groups based on their DNA and amino acid sequences (Bray, 1994). However, the *in vivo* functions of these genes during adaptation or tolerance to water deficit have not yet been clearly identified, although recent results (Xu et al., 1996) suggest that this may be the case.

Cold acclimation is accompanied by alterations in protein synthesis (Darly et al., 1995) and in membrane lipid composition, including an increase of trienoic fatty acid (Kodama et al., 1994; 1995; Uemura et al., 1995). Several lines of evidence suggest that ABA plays a role in cold acclimation. For example, exposure to exogenous ABA can induce freezing tolerance in plants at normal temperature (Mäntylä et al., 1995), and the level of endogenous ABA is elevated during cold acclimation (Lång et al., 1994). A number of cold-regulated genes have been isolated from *Arabidopsis* and barley; these genes encode proteins such as dehydrin-like proteins (van Zee et al., 1995), CORs (cold regulated proteins) and KINs (kykna-indusoitu proteins) (reviewed in Thomashow, 1994). Like drought-inducible genes, most of these are also induced by exogenous ABA as well as by drought stress.

Furthermore, a number of genes induced by ABA application are also expressed in response to salt. A *RAB* gene (*16 A*) in rice (Mundy and Chua, 1988), *SALT* in rice (Claes et al., 1990), a kinase homologue in *Arabidopsis* (Hwang and Goodman, 1995), and *TAS14* in tomato (Godoy et al., 1990) are expressed during salt acclimation. *RD29A* and *RD29B* genes were isolated from *Arabidopsis* under drought stress; these genes were also induced by salt, low temperature, and exogenous ABA (Yamaguchi-Shinozaki and Shinozaki, 1994).

ABA-responsive cis *regulatory elements are also diverse*

The mechanisms by which ABA regulates gene expression involve transcriptional and post-transcriptional events. Here, we will focus on the ABA-mediated transcriptional control of gene expression. Marcotte et al. (1989) first identified an ABA responsive element (ABRE) in the wheat Em gene; this element is similar to the G-box motifs found in many lights, including UV-light, and anaerobiosis induced genes (Mikami et al., 1995; Schindler et al., 1992; Weisshaar et al., 1991; Staiger et al., 1989; Giuliano et al., 1988). A number of *cis*-elements containing the consensus ABRE, CACGTG were identified in many ABA-responsive genes including cold-, and drought-inducible genes either by functional analysis or by sequence similarity; most functionally defined ABREs contained the core sequence, ACGT (Tab. 1). However, promoter deletion and mutation analysis of several ABA-responsive genes subsequently demonstrated that not all of these genes contain the consensus ABRE (Tab. 1). Furthermore, some Em-like ABREs are either not functional or are insufficient for ABA response (see Tab. 1). Dehydration-responsive genes, *rd22* and *rd29* are ABA inducible, but do not contain any *cis* elements related to the Em-like ABRE, but rather contain an element for drought response, DRE, in their promoters (Yamaguchi-Schinozaki and Shinozaki, 1994; Iwasaki et al., 1995). Four putative ABREs in the gene *CdeT27-45* of the resurrection plant did not respond to exogenous ABA (Michal et al., 1993), whereas the sequences containing two repeats of AGCCCA were identified by DNase I footprinting, and its DNA-binding activity was ABA-inducible (Nelson et al., 1994). Furthermore, Straub et al. (1994) reported that the consensus ABREs in the barley *HVA1* are not sufficient for ABA induced expression. Similar results obtained with the *cor15a* (Baker et al., 1994) and *Dc3* genes (see below) support the above observations.

The carrot gene *Dc3* is representative of the variation found in ABA-responsive genes and their cognate *cis* regulatory elements. *Dc3* is a *LEA* class gene and is normally expressed in developing seeds and in vegetative tissues in response to drought and exogenous ABA like other *LEA* genes (Seffens et al., 1990; Vivekananda et al., 1992). The *Dc3* promoter consists of two distinct regulatory regions (Fig. 1). The distal promoter region (DPR) confers ABA responsive expression in vegetative tissues; whereas the proximal promoter region (PPR) confers seed-specific expression and does not respond to ABA (see Thomas, 1993). To identify the *cis*-elements involved in ABA response, various 5' upstream deletions, internal deletions and site-directed mutations were constructed within the context of the PPR (Fig. 1). These constructs were fused to the GUS reporter gene, transferred to tobacco and assayed for GUS expression by exposure to ABA or under drought stress in vegetative tissues. Although the DPR contains four ACGT-containing sequences between -807 to -235, three of these have been shown not to respond to

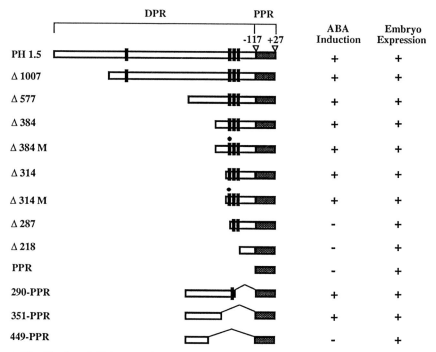

Figure 1. Identification of ABA-responsive DNA sequences in the *Dc3* promoter. PH 1.5 is the full-length *Dc3* gene promoter; it is composed of two distinct regulatory regions: DPR (distal promoter region) and PPR (proximal promoter region). Small vertical rectangles represent ACGT-containing sequences located at -807, -300, -256, and -235. The small solid circle indicates the ACGT-containing sequence at -300 mutated from TACGTGTC to TGAGATCT. Seeds from transgenic tobacco (R1) containing deletions and mutations of the *Dc3* promoter fused to the β-glucuronidase (GUS) reporter gene were germinated on kanamycin selection media, and 10 day seedlings were incubated in MS liquid media with 100 μM ABA. GUS activity was determined after 1 or 4 days treatment. Constructs that respond to ABA with an induction greater than twice that of the corresponding control are indicated by a +.

exogenous ABA as determined by deletion and mutagenesis experiments, implying that the last ACGT-containing sequence at -807 is not necessary for ABA induction since further downstream deletions respond to ABA (Fig. 1). However, the regions between -577 to -351 or -577 to -290 fused to the *Dc3* PPR do respond to ABA, but the region between -577 to -449 does not, indicating the sequences between -449 to -351 and -314 to -301 contain the *cis*-elements that confer ABA inducible expression in the context of the PPR and the ACGT sequence does not. In addition, these same regions fused to the -90 truncated CaMV 35S promoter were not induced by exogenous ABA in vegetative tissues (data not shown). These results show that the PPR is necessary for ABA induction even though the PPR itself is ABA-independent and that combinatorial interactions of *cis*-elements, located between the PPR and DPR, with their cognate *trans*-acting factors occur during the response to ABA. Interestingly, there is no Em-like ABRE in this region as found in many ABA-inducible genes, instead, TTTCGTGT motifs are repeated several times between -449 to -301. We have now determined that the *trans*-acting factors that specifically bind to the upstream regions are different from the ones that bind to the PPR. Furthermore, recent results indicate the TTTCGTGT motif is a functional ABRE in vegetative tissues.

The putative ABREs of two helianthinin genes encoding sunflower legumin seed storage proteins also fall into a class of distantly related Em consensus sequences. The helianthinin gene is normally expressed exclusively in the developing seed, but its expression can be modulated by the application of ABA (Bogue et al., 1990; Nunberg et al., 1994). Based on sequence comparison, the helianthin ABRE is TACGAACC which shares less than 75% similarity with the consensus CACGTGGC. Proof of function for the putative helianthinin ABREs has involved gain-of-function experiments with overlapping 5' regions from two helianthinin genes, *HaG3-A* and *HaG3-D*, fused to a truncated CaMV 35S promoter. It is noteworthy that the helianthinin ABREs, one in *HaG3-D* and two in *HaG3-A*, are adjacent to AT-rich sequences that bind HMG-like proteins (J. Jordano and T. Thomas, unpublished results).

The simplest and most straight forward conclusion to be drawn from Table 1 is that there are multiple ABREs with different consensus DNA sequences; in some cases, additional components may be required for proper ABA response. Rogers and Rogers (1992) suggested that coupling elements are required for response to ABA as well as GA. In this study, the *cis*-acting element O2S is similar to the binding site for *Opaque-2* that functions as a coupling element, resulting in enhancement of hormone-regulated transcription with the ABRE or the GARE within the context of the Amy32B promoter (Rogers and Rogers, 1992). In addition to O2S, Shen and Ho (1995) found another element CE1 (TGCCACCGG) in the promoter of the barley *HVA22* gene that couples with a G-box (GCCACGTACA) element to constitute a bipartite ABRE that confers high-levels of ABA induction. The corresponding *trans*-acting factors for these elements have not been identified, However, the transcriptional activator VP1 has been identified, and it might be a good candidate for this role. VP1 is required for induction of the maize Em homolog during embryo development (McCarty et al., 1991). VP1 and ABA function synergistically to activate GUS reporter gene expression driven by chimeric promoters containing tetramers of either the Em1a or the Em1b G-box regions and the proximal (-45) CaMV 35S promoter fused in turn to the GUS reporter gene (Vasil et al., 1995). Disruption of both Em1a and Em1b elements indeed does have an effect on VP1 synergism, indicating that VP1 may interact with other *cis*-elements, possibly the coupling element *via* protein-protein interaction. Similarly, Hattori et al. (1995) showed that a rice VP1 homologue also activated the expression of the rice Em gene, *Osem*, with exogenous ABA. Mutational analysis using a rice protoplasts expression system revealed that the coupling element CGGCGGCCTCGCCACG (region1) is adjacent to the ABRE (TACGTGTC) in the rice *Osem* promoter and that this sequence was important for VP1 activation. Furthermore, Bobb et al. (1995) cloned a bean (*P. vulgaris*) homologue of *VP1*, *PvAlf*; transient expression of *PVALF* in protoplasts containing a phaseolin promoter GUS transgene resulted in GUS expression.

The nuclear proteins EMBP-1 and TAF-1 belong to the bZIP family of transcription factors and were cloned from wheat and tobacco, respectively. EMBP-1 specifically binds to the Em1a motif, GACACGTGGC in the wheat Em gene, and TAF-1 binds to motif 1, GTACGTGGCG, of the rice *RAB* gene promoter (Oeda et al., 1991; Guiltinan et al., 1990). However, since the identification of EMBP-1 and TAF-1, other ABRE binding factors have not been cloned, whereas genes encoding many of other bZIP proteins which bind to the G-box or G-box-like (GBL) sequences of several gene classes have been cloned (de Vetten and Ferl, 1995; Hong et al., 1995; Mikami et al., 1995; Izawa et al., 1993). This raises two questions: What are the differences between the G-box and the prototypical ABREs, and what determines the specificity of the transcription factors that bind to these ABREs, but not to the G-box? There might be unidentified transcription factors that are distant bZIP family members or that do not belong to the bZIP family at all and are only involved in the ABA response. For example, the ABREs of the rice *RAB-16 A* and the resurrection plant gene *CDeT27-45* are bound by nuclear proteins from both control and ABA-treated tissues (Mundy et al., 1990; Nelson et al., 1994). In addition, nuclear extracts from ABA-treated tissues do not bind preferentially to sequences conferring ABA inducibility in the Dc3 DPR (H. Chung and T. Thomas, unpublished results). The drought response element (DRE) in *rd29A* is another example of an element that responds to drought, low-temperature, and salt, yet does not bind differentially to nuclear extracts prepared from tissues exposed to these stimuli (Yamaguchi-Shinozaki and Shinozaki, 1994). If the hypothesis of a coupling element for ABA induction is correct, it is possible that the ABREs and their cognate transcription factors may be expressed constitutively and activation *via* ABA requires: (1) the interaction, directly or indirectly, of unidentified *trans*-acting factors, perhaps like VP1, with coupling elements, (2) post-translational modifications of the ABRE binding factor(s), or (3) the formation of heterodimers with other nuclear proteins resulting in an active ABA-responsive complex.

Conclusions

From the above, it is clear that ABA is part of a signal transduction pathway by which plants respond to developmental cues and adapt to changes in environmental conditions. However, the signal transduction pathway is not rectilinear. Thus, while ABA signaling is important for the proper maturation of the embryo, it is not the only developmental program involved in this process. Similarly, ABA is clearly involved in many aspects of the plant's response to drought, salt and cold stress, but other signal transduction pathways are also important. Furthermore, the diversity of the genes and *cis* regulatory elements that respond to ABA suggests a corresponding diversity in their intersection with the two or more ABA signaling networks currently proposed.

There is an abundance of phenomenological data concerning the physiological and molecular events associated with ABA perception and signal transduction. The tools are now in place to address this complex problem. However, progress has been modest, so far, in applying molecular genetic approaches to elucidate the cellular and molecular events required for ABA mediated signal transduction. Although there are mutations that involve ABA synthesis and perception in several plants, *Arabidopsis* is the plant of choice to continue these studies. Future work should focus on mutations involved in ABA perception, comprising at the moment the *abi* mutant series. Additional studies are needed to examine the interaction of the increasing number of *abi* alleles with other genes either directly or by using appropriate reporter/promoter constructions. In addition, it is not clear that all ABA perception loci have been isolated, thus an additional mutagenesis and screen is warranted, perhaps using a transgenic line containing an ABA responsive promoter driving expression of a screenable reporter such as GUS or the green fluorescent protein (GFP). The combination of single cell technologies and transgenic plants containing ABA responsive promoters fused to sensitive reporter genes like GFP and GUS should facilitate analysis of the role of second messengers in ABA signaling. Furthermore, an important next step in the dissection of the *Arabidopsis* signal transduction pathway is to conduct a suppressor screen starting with weak and strong alleles of each *abi* locus. Results from the two types of screens should produce additional loci to further elucidate the ABA mediated signal transduction pathway.

Acknowledgements
Work from the author's lab was supported by grants from the U.S. Department of Agriculture (9493199) and Rhône Poulenc Agrochemie.

References

Allan, A.C., Fricker, M.D., Ward, J.L., Beale, M.H. and Trewavas, A.J. (1994) Two transduction pathways mediate rapid effects of abscisic acid in *Commelina* guard cells. *Plant Cell* 6: 1319–1328.
Anderberg, R.J. and Walker-Simmons, M.K. (1992) Isolation of a wheat cDNA clone for an abscisic acid-inducible transcript with homology to protein kinases. *Proc. Natl. Acad. Sci. USA* 88: 8602–8605.
Anderson, B.E., Ward, J.M. and Schroeder, J.I. (1994) Evidence for an extracellular reception site for abscisic acid in *Commelina* guard cells. *Plant Physiol.* 104: 1177–1183.

Armstrong, F., Leung, J., Grabov, A., Brearley, J., Giraudat, J. and Blatt, M.R. (1995) Sensitivity to abscisic acid of guard-cell K+ channels is suppressed by *abi1-1*, a mutant *Arabidopsis* gene encoding a putative protein phosphatase. *Proc. Natl. Acad. Sci. USA* 92: 9520–9524.

Baker, S.S., Wilhelm, K.S. and Thomashow, M.F. (1994) The 5'-region of *Arabidopsis thaliana cor15a* has *cis*-acting elements that confer cold-, drought- and ABA-regulated gene expression. *Plant Molec. Biol.* 24: 701–713.

Bäumlein, H., Miséra, S., Luersen, H., Koelle, K., Horstmann, C., Wobus, U. and Mueller, A.J. (1994) The *FUS3* gene of *Arabidopsis thaliana* is a regulator of gene expression during late embryogenesis. *Plant J.* 6: 379–387.

Black, M. (1991) Involvement of ABA in the physiology of developing and mature seeds. *In*: W.J. Davies and H.G. Jones (eds): *Abscisic Acid: Physiology and Biochemistry*. Bios, Oxford, pp 99–124.

Blatt, M.R. and Thiel, G. (1993) Hormonal control of ion channel gating. *Annu. Rev. Plant Physiol.* 44: 543–67.

Bobb, A.J., Eiben, H.G. and Bustos, M.M. (1995) PvAlf, an embryo-specific acidic transcriptional activator enhances gene expression from phaseolin and phytohemagglutinin promoters. *Plant J.* 8: 331–43.

Bogue, M.A., Vonder Haar, R.A., Nuccio, M.L., Griffing, L.R. and Thomas, T.L. (1990) Developmentally regulated expression of a sunflower 11S seed protein gene in transgenic tobacco. *Molec. Gen. Genet.* 221: 49–57.

Bray, E.A. (1991) Regulation of gene expression by endogenous ABA during drought stress. *In*: W.J. Davies and H.G. Jones (eds): *Abscisic Acid: Physiology and Biochemistry*. Bios, Oxford, pp 79–97.

Bray, E.A. (1994) Alterations in gene expression in response to water deficit. *In*: A.S. Basra (ed.): *Stress-Induced Gene Expression in Plants*. Harwood Academic Publishers, Chur (India), pp 1–23.

Briskin, D.P. and Hanson, J.B. (1992) How does the plant plasma membrane H+-ATPase pump protons? *J. Exp. Bot.* 43: 269–289.

Chapman, K.S.R., Trewavas, A. and van Loon, L.C. (1975) Regulation of the phosphorylation of chromatin-associated proteins in *Lemna* and *Hordeum*. *Plant Physiol.* 55: 293–296.

Claes, B., Dekeyser, R., Villarroel, R., Van den Bulcke, M., Bauw, G., Van Montagu, M. and Caplan, A. (1990) Characterization of a rice gene showing organ-specific expression in response to salt stress and drought. *Plant Cell* 2: 19–27.

Covarrubias, A.A., Ayala, J.W., Reyes, J.L., Hernandez, M. and Garciarrubio, A. (1995) Cell-wall proteins induced by water deficit in bean (*Phaseolus vulgaris* L.) seedlings. *Plant Physiol.* 107: 1119–1128.

Creelman, R.A., Mason, H.S., Benson, R.J., Boyer, J.S. and Mullet, J.E. (1990) Water deficit and abscisic acid cause differential inhibition of shoot *versus* root growth in soybean seedlings. *Plant Physiol.* 92: 205–214.

Darley, C.P., Davies, J.M. and Sanders, D. (1995) Chill-induced changes in the activity and abundance of the vacuolar proton-pumping pyrophosphatase from mung bean hypocotyls. *Plant Physiol.* 109: 659–665.

de Vetten, N.C. and Ferl, R.J. (1995) Characterization of a maize G-box binding factor that is induced by hypoxia. *Plant J.* 7: 589–601.

Dure, L.,III. Crouch, M., Harada, J., Ho, T.H.D., Mundy, J., Quatrano, R., Thomas, T. and Sung, Z.R. (1989) Common amino acid sequence domains among the LEA proteins of higher plants. *Plant Molec. Biol.* 12: 475–486.

Dure, L., III. (1993) The Lea proteins of higher plants. *In*: D.P.S.Verma (ed.): *Control of Plant Gene Expression*, CRC Press, Boca Raton, pp 325–335.

Finkelstein, R. (1994) Mutations at two new *Arabidopsis* ABA response loci are similar to ABI3 mutations. *Plant J.* 5: 765–771.

Finkelstein, R. and Somerville, C.R. (1990) Three classes of abscisic acid (ABA)-insensitive mutations of *Arabidopsis* define genes that control overlapping subsets of ABA responses. *Plant Physiol.* 94: 1172–1179.

Gilroy, S. and Jones, R.L. (1994) Perception of gibberellin and abscisic acid at the external face of the plasma membrane of barley (*Hordeum vulgare* L.) aleurone protoplasts. *Plant Physiol.* 104: 1185–1192.

Gilroy, S., Read, N.D. and Trewavas, A.J. (1990) Elevation of cytoplasmic calcium by caged calcium or caged inositol triphosphate initiates stomatal closure. *Nature* 346: 769–771.

Giuliano, G., Pichersky, E., Malik, V.S., Timko, M.P. Scolnik, P.A. and Cashmore, A.R. (1988) An evolutionary conserved protein binding sequence upstream of a plant light-regulated gene. *Proc. Natl. Acad. Sci. USA* 85: 7089–7093.

Giraudat, J. (1995) Abscisic acid signaling. *Curr. Opin. Cell Biol.* 7: 232–238.

Giraudat, J., Parcy, F., Bertauche, N., Gosti, F., Leung, J., Morris. P-C., Bouvier-Durand, M. and Vartanian, N. (1994) Current advances in abscisic acid signaling. *Plant Molec. Biol.* 26: 1557–1577.

Godoy, J.A., Pardo, J.M. and Pintor-Toro, J.A. (1990) A tomato cDNA inducible by salt stress and abscisic acid: nucleotide sequence and expression pattern. *Plant Molec. Biol.* 15: 695–705.

Gosti, F., Bertauche, N., Vartanian, N. and Giraudat, J. (1995) Abscisic acid-dependent and -independent regulation of gene expression by progressive drought in *Arabidopsis thaliana*. *Molec. Gen. Genet.* 246: 10–18.

Guiltinan, M.J., Marcotte Jr. W.R. and Quatrano, R.S. (1990) A plant leucine zipper protein that recognizes an abscisic acid response element. *Science* 250: 267–271.

Hattori, T., Vasil, V., Rosenkrans, L., Hannah, L.C., McCarty, D.R. and Vasil, I.K. (1992) The *Viviparous-1* gene and abscisic acid activate the *C1* regulatory gene for anthocyanin biosynthesis during seed maturation in maize. *Genes Dev.* 6: 609–618.

Hattori, T., Terada, T. and Hamasuna, S. (1994) Sequence and functional analyses of the rice gene homologous to the maize *Vp1*. *Plant Molec. Biol.* 24: 805–810.

Hattori, T., Terada, T. and Hamasuna, S. (1995) Regulation of the *Osem* gene by abscisic acid and the transcriptional activator VP1: analysis of *cis*-acting promoter elements required for regulation by abscisic acid and VP1. *Plant J.* 7: 913–925.

Hedrich, R. and Schroeder, J.I. (1989) The physiology of ion channels and electrogenic pumps in higher plants. *Annu. Rev. Plant Physiol.* 40: 539–569.

Heino, P., Samdman, G., Lang, V., Nordin, K. and Palva, E.T. (1990) Abscisic acid deficiency prevents development of freezing tolerance in *Arabidopsis thaliana* (L.) Heynh. *Theoret. Appl. Genet.* 79: 801–806.

Hoecker, V., Vasil, I.K. and McCarty, D.R. (1995) Integrated control of seed maturation and germination programs by activator and repressor functions of *viviparous-1*. *Genes Dev.* 9: 2459–2469.

Hong, J.C., Cheong, Y.H., Nagao, R.T., Bahk, J.D., Key, J.L. and Cho, M.J. (1995) Isolation of two soybean G-box binding factors which interact with a G-box sequence of an auxin-responsive Gene. *Plant J.* 8: 199–211.

Hornberg, C. and Weiler, E.W. (1984) High-affinity binding sites for abscisic acid on the plasmalemma of *Vicia faba* guard cells. *Nature* 310: 321–324.

Hwang, I. and Goodman, H.M. (1995) An *Arabidopsis thaliana* root-specific kinase homolog is induced by dehydration, ABA, and NaCl. *Plant J.* 8: 37–43.

Irving, H.R., Gehring, C.A. and Parish, R.W. (1992) Changes in cytosolic pH and calcium of guard cells precede stomatal movements. *Proc. Natl. Acad. Sci. USA* 89: 1790–1794.

Iwasaki, T., Yamaguchi-Shinozaki, K. and Shinozaki, K. (1995) Identification of a *cis*-regulatory region of a gene in *Arabidopsis thaliana* whose induction by dehydration is mediated by abscisic acid and requires protein synthesis. *Molec. Gen. Genet.* 247: 391–398.

Izawa, T., Foster, R. and Chua, N.-H. (1993) Plant bZIP protein DNA binding specificity. *J. Molec. Biol.* 230: 1131–1144.

Kaiser, W.M and Hartung, W. (1981) Uptake and release of abscisic acid by isolated photoautotrophic mesophyll cells, depending on pH gradients. *Plant Physiol.* 68: 202–206.

Keith, K., Kraml, M., Dengler, N.G. and McCourt, P. (1994) *fusca3*: A heterochronic mutation affecting late embryo development in *Arabidopsis*. *Plant Cell* 6: 589–600.

Kinoshita, T., Mitsuo, N. and Shimazaki, K.-I. (1995) Cytosolic concentration of Ca^{2+} regulates the plasma membrane H^+-ATPase in guard cells of fava bean. *Plant Cell* 7: 1333–1342.

Kodama, H., Hamada, T., Horiguchi, G., Nishimura, M. and Iba, K. (1994) Genetic enhancement of cold tolerance by expression of a gene for chloroplast ω-3 fatty acid desaturase in transgeneic tobacco. *Plant Physiol.* 105: 601–605.

Kodama, H., Horiguchi, G., Nishiuchi, T., Nishimura, M. and Iba, K. (1995) Fatty acid desaturation during chilling acclimation is one of the factors involved in conferring low-temperature tolerance to young tobacco leaves. *Plant Physiol.* 107: 1177–1185.

Koontz, D.A. and Choi, J.H. (1993) Protein phosphorylation in carrot somatic embryos in response to abscisic acid. *Plant Physiol. Biochem.* 31: 95–102.

Koornneef, M., Jorna, M.L., Brinkhorst-van der Swan, D.L.C. and Karssen, C.M. (1982) The isolation of abscisic acid deficient mutants by selection of induced revertants in non-germinating gibberellin sensitive lines of *Arabidopsis thaliana* (L.) Heynh. *Theoret. Appl. Genet.* 61: 385–393.

Koornneef, M., Reuling, G. and Karssen, C.M. (1984) The isolation and characterization of abscisic acid-insensitive mutants of *Arabidopsis thaliana*. *Plant. Physiol.* 61: 377–383.

Lång, V. and Palva, E.T. (1992) The expression of a rab-related gene, rab18, induced by abscisic acid during the cold acclimation process of *Arabidopsis thaliana* (L.) Heynh. *Plant Molec. Biol.* 20: 951–962.

Lång, V., Mäntylä, E., Welin, B., Sundberg, B. and Palva, E.T. (1994) Alteration in water status, endogenous ABA content, and expression of *rab18* gene during the development of freezing tolerance in *Arabidopsis thaliana*. *Plant Physiol.* 104: 1341–1349.

Leung, J., Bouvier-Durand, M., Morris, P.-C., Guerrier, D., Chefdor, F. and Giraudat, J. (1994) *Arabidopsis* ABA response gene *ABI1*: features of a calcium-modulated protein phosphatase. *Science* 264: 1448–1452.

MacRobbie, E.A.C. (1995) ABA-induced ion efflux in stomatal guard cells: Multiple actions of ABA inside and outside the cell. *Plant J.* 7: 565–576.

Mäntylä, E., Lång, V. and Palva, E.T. (1995) Role of abscisic acid in drought-induced freezing tolerance, cold acclimation, and accumulation of LTI78 and RAB18 proteins in *Arabidopsis thaliana*. *Plant Physiol.* 107: 141–148.

Marcotte, W.R., Russel, S.H. and Quatrano, R.S. (1989) Abscisic acid-responsive sequence from the Em gene of wheat. *Plant Cell* 1: 969–976.

McCarty, D.R. (1995) Genetic control and integration of maturation and germination pathways in seed development. *Annu. Rev. Plant Physiol.* 46: 71–93.

McCarty, D.R., Hattori, T., Carson, C.B., Vasil, V., Lazar, M. and Vasil, I.K. (1991) The *Viviparous-1* developmental gene of maize encodes a novel transcriptional activator. *Cell* 66: 895–905.

Meinke, D.W, Franzmann, L.H., Nickle, T.C. and Yeung, E.C. (1994) *Leafy cotyledon* mutants of *Arabidopsis*. *Plant Cell* 6: 1049–1064.

Meurs, C., Basra, A., Karseen, C.M. and van Loon, L.C. (1992) Role of abscisic acid in the induction of desiccation tolerance in developing seeds of *Arabidopsis thaliana*. *Plant Physiol.* 98: 1484–1493.

Michel, D., Salamini, F., Bartels, D., Dale, P., Baga, M. and Szalay, A. (1993) Analysis of a desiccation and ABA-responsive promoter from the resurrection plant *Craterostigma plantagineum*. *Plant J.* 4: 29–40.

Michel, D., Furini, A., Salamini, F. and Bartels, D. (1994) Structure and regulation of an ABA- and desiccation-responsive gene from the resurrection plant *Craterostigma plantagineum*. *Plant Molec. Biol.* 24: 549–560.

Mikami, K., Katsura, M., Ito, T., Okada, K., Shimura, Y. and Iwabuchi, M. (1995) Developmental and tissue-specific regulation of the gene for the wheat basic/leucine zipper protein HBP-1a(17) in transgenic *Arabidopsis* plants. *Molec. Gen. Genet.* 248: 573–582.

Mundy, J. and Chua, N-.H. 1988. Abscisic acid and water-stress induce the expression of a novel rice gene. *EMBO J.* 7: 2279–2286.

Mundy, J., Yamaguchi-Shinozaki, K. and Chua, N-.H. (1990) Nuclear proteins bind conserved elements in the abscisic acid-responsive promoter of rice rab gene. *Proc. Natl. Acad. Sci. USA* 87: 1046–1410.

Naito, S., Hirai, M.Y., Chino, M. and Komeda, Y. (1994) Expression of a soybean (*Glycine max* [L.] Merr.) seed storage protein gene in transgenic *Arabidopsis thaliana* and its response to nutritional stress and to abscisic acid mutations. *Plant Physiol.* 104: 497–503.

Nambara, E., Naito, S. and McCourt, P. (1992) A mutant of *Arabidopsis* which is defective in seed development and storage protein accumulation is a new *abi3* allele. *Plant J.* 2: 435–441.

Nambara, E., Keith, K., McCourt, P. and Naito, S. (1994) Isolation of an internal deletion mutant of the *Arabidopsis thaliana ABI3* gene. *Plant Cell Physiol.* 35: 509–513.

Nambara, E., Keith, K., McCourt, P. and Naito, S. (1995) A regulatory role for the *ABI3* gene in the establishment of embryo maturation in *Arabidopsis thaliana*. *Devel.* 121: 629–636.

Nelson, D., Salamini, F. and Bartels, D. (1994) Abscisic acid promotes novel DNA-binding activity to a desiccation-related promoter of *Craterostigma plantagineum*. *Plant J.* 5: 451–458.

Nordin, K., Vahala, T. and Palva, E.T. (1993) Differential expression of two related, low-temperature-induced genes in *Arabidopsis thaliana* (L.) Heynh. *Plant Molec. Biol.* 21: 641–653.

Nunberg, A.N., Li, Z.W., Bogue, M.A., Vivekananda, J., Reddy, A.S. and Thomas, T.L. (1994) Developmental and hormonal regulation of sunflower helianthinin genes: Proximal promoter sequences confer regionalized seed expression. *Plant Cell* 6: 473–486.

Oeda, K., Salinas, J. and Chua, N.-H. (1991) A tobacco bZIP transcription activator (TAF-1) binds to a G-box-like motif conserved in plant genes. *EMBO J.* 10: 1793–1802.

Ooms, J.J.J., Léon-Kloosterziel, K.M., Bartels, D., Koornneef, M. and Karssen, C.M. (1993) Acquisition of desiccation tolerance and longevity in seeds of *Arabidopsis thaliana*. *Plant Physiol.* 102: 1185–1191.

Paiva, R. and Kriz, A.L. (1994) Effect of abscisic acid on embryo-specific gene expression during normal and precocious germination in normal and *viviparous* maize (*Zea mays*) embryos. *Planta* 192: 332–339.

Parcy, F., Valon, C., Raynal, M., Gaubier-Comella, P., Delseny, M. and Giraudat, J. (1994) Regulation of gene expression programs during *Arabidopsis* seed development: Roles of the *ABI3* locus and of endogenous abscisic acid. *Plant Cell* 6: 1567–1582.

Parmar, P.N. and Brearley, C.A. (1993) Identification of 3- and 4-phosphorylated phosphoinositides and inositol phosphates in stomatal guard cells. *Plant J.* 4: 255–263.

Phillips, J. and Conrad, U. (1994) Genomic sequences from the *Nicotiana tabacum* homologous to the maize transcriptional activator gene *Viviparous-1*. *J. Plant Physiol.* 144: 760–761.

Pla, M., Vilardell, J., Guiltinan, M.J., Marcotte, W.R., Niogret, M.F. Quatrano, R.S. and Pagès, M. (1993) The *cis*-regulatory element CCACGTGG is involved in ABA and water-stress responses of the maize gene *rab28*. *Plant Molec. Biol.* 21: 259–266.

Plant, A.L., van Rooijen, G.J.H., Anderson, C.P. and Moloney, M.M. (1994) Regulation of an *Arabidopsis* oleosin gene promoter in transgenic *Brassica napus*. *Plant Molec. Biol.* 25: 193–205.

Quatrano, R.S., Marcotte, Jr. W.R. and Guiltinan, M. (1993) Regulation of gene expression by abscisic acid. *In*: D.P.S. Verma (ed.): *Control of Plant Gene Expression*. CRC Press, Boca Raton, pp 69–90.

Reviron, M.-P., Vartanian, N., Sallantin, M., Huet, J.-C., Pernollet, J.-C. and de Vienne, D. (1992) Characterization of a novel protein induced by progressive or rapid drought and salinity in *Brassica napus* leaves. *Plant Physiol.* 100: 1486–1493.

Rogers, J.C. and Rogers, S.W. (1992) Definition and functional implications of gibberellin and abscisic acid *cis*-acting hormone response complexes. *Plant Cell* 4: 1443–1451.

Schindler, U., Terzaghi, W., Beckmann, H., Kadesch, T. and Cashmore, A.R. (1992) DNA binding site preferences and transcriptional activation properties of the *Arabidopsis* transcription factor GBF1. *EMBO J.* 11: 1275–1289.

Schmidt, C., Schelle, I., Liao, Y.H. and Schroeder J.I. (1995) Strong regulation of slow anion channels and abscisic acid signaling in guard cells by phosphorylation and dephosphorylation events. *Proc. Natl. Acad. Sci. USA* 92: 9535–9539.

Schroeder, J.I. and Hagiwara, S. (1989) Cytosolic calcium regulates ion channels in the plasma membrane of *Vicia faba* guard cells. *Nature* 338: 427–430.

Schroeder, J.I. and Hagiwara, S. (1990) Repetitive increases in cytosolic Ca^{2+} of guard cells by abscisic acid activation of nonselective Ca^{2+} permeable channels. *Proc. Natl. Acad. Sci. USA* 87: 9305–9309.

Schroeder, J.I. and Keller, B.U. (1992) Two types of anion channel currents in guard cells with distinct voltage regulation. *Proc. Natl. Acad. Sci. USA* 89: 5025–5029.

Schroeder, J.I., Schmidt. C. and Sheaffer, J. (1993) Identification of high-affinity slow anion channel blockers and evidence for stomatal regulation by slow anion channels in guard cells. *Plant Cell* 5: 1831–1841.

Schwartz, A., Wu, W.H., Tucker, E.B. and Assmann, S.M. (1994) Inhibition of inward K$^+$ channels and stomatal response by abscisic acid – an intracellular locus of phytohormone action. *Proc. Natl. Acad. Sci. USA* 91: 4019–4023.

Schwartz, A., Ilan, N., Schwartz, M., Scheaffer, J., Assmann, S. and Schroeder, J.I. (1995) Anion-channel blockers inhibit S-type anion channels and abscisic acid responses in guard cells. *Plant Physiol.* 109: 651–658.

Seffens, W.S., Almoguera, D., Wilde, H.D., Vonder Haar, R.A. and Thomas, T.L. (1990) Molecular analysis of a phylogenetically conserved carrot gene: Developmental and environmental regulation. *Dev. Genet* 11: 65–76.

Shen, Q. and Ho, T.-H.D. (1995) Functional dissection of an abscisic acid(ABA)-inducible gene reveals two independent ABA-responsive complexes each containing a G-box and a novel *cis*-acting element. *Plant Cell* 7: 295–307.

Skriver, K. and Mundy, J. (1990) Gene expression in response to abscisic acid and osmotic stress. *Plant Cell* 2: 503–512.

Staiger, D., Kaulen, H. and Schell, J. (1989) A CACGTG motif of the *Anthirrhinum majus* chalcone synthase promoter is recognized by an evolutionarily conserved nuclear protein. *Proc. Natl. Acad. Sci. USA* 86: 6930–6934.

Straub, P.F., Shen, Q. and Ho, T.-H.D. (1994) Structure and promoter analysis of an ABA- and stress-regulated barley gene, *HVA1*. *Plant Molec. Biol.* 26: 617–630.

Taylor, J.E., Renwick, K., Webb, A.A.R., McAinsh, M.R., Furini, A., Bartels, D., Quatrano, R.S., Marcotte, W.R. and Hetherington, A.M. (1995) ABA-regulated promoter activity in stomatal guard cells. *Plant J.* 7: 129–134.

Tazawa, M., Shimmen, T. and Mimura, T. (1987) Membrane control in the characeae. *Annu. Rev. Plant Physiol.* 38: 95–117.

Thiel, G., MacRobbie, E.A.C. and Blatt, M.R. (1992) Membrane transport in stomatal guard cells: the importance of voltage control. *J. Memb. Biol.* 126: 1–18.

Thomas, T.L. (1993) Gene expression during plant embryogenesis and germination: An overview. *Plant Cell* 5: 1401–1410.

Thomashow, M.F. (1994) *Arabidopsis thaliana* as a model for studying mechanisms of plant cold tolerance. *In:* C.R. Somerville and E.M. Meyerowitz (eds): *Arabidopsis*, CSH Press, New York, pp 807–834.

Uemura, M., Joseph, R.A. and Steponkus, P.L. (1995) Cold acclimation of *Arabidopsis thaliana*. Effect on plasma membrane lipid composition and freeze-induced lesions. *Plant Physiol.* 109: 15–30.

van Zee, K., Chen, F.Q., Hayes, P.M., Close, T.J. and Chen, T.H.H. (1995) Cold-specific induction of a dehydrin gene family member in barley. *Plant Physiol.* 108: 1233–1239.

Vasil, V., Marcotte, Jr. W.R., Rosenkrans, L., Cocciolone, S.M., Vasil, I.K., Quatrano, R.S. and McCarty, D.R. (1995) Overlap of viviparous (VP1) and abscisic acid response elements in the *Em* promoter: G-box elements are sufficient but not necessary for VP1 transactivation. *Plant Cell* 7: 1511–1518.

Vilardell, J., Mundy, J., Stilling, B., Leroux, B., Pla, M., Freyssinet, G. and Pagès, M. (1991) Regulation of the maize rab17 gene promoter in transgenic heterologous systems. *Plant Molec. Biol.* 17: 985–993.

Vivekananda, J., Drew, M.C. and Thomas, T.L. (1992) Hormonal and environmental regulation of the carrot *lea*-class gene *Dc3*. *Plant Physiol.* 100: 576–581.

Wang, H., Datla, R., Georges, F., Loewen, M. and Cutler, A.J. (1995) Promoters from *kin1* and *cor6.6*, two homologous *Arabidopsis thaliana* genes: transcriptional regulation and gene expression induced by low temperature, ABA, osmoticum and dehydration. *Plant Molec. Biol.* 28: 605–617.

Weisshaar, B., Armstrong, G.A., Block, A., da Costa e Silva, O. and Hahlbrock, K. (1991) Light-inducible and constitutively expressed DNA-binding proteins recognizing a plant promoter element with functional relevance in light responsiveness. *EMBO J.* 10: 1777–1786.

Wu, Y., Spollen, W.G., Sharp, R.E., Hetherington, P.R. and Fry, S.C. (1994) Root growth maintenance at low water potentials. *Plant Physiol.* 106: 607–615.

Xu, D., Duan, X., Wang, B., Hong, B., Ho, T.-H.D. and Wu, R. (1996) Expression of a late embryogenesis abundant protein gene, HVA1, from barley confers tolerance to water deficit and salt stress in transgenic rice. *Plant Physiol.* 110: 249–57.

Yamaguchi-Shinozaki, K. and Shinozaki, K. (1994) A novel *cis*-acting element in an *Arabidopsis* gene is involved in responsiveness to drought, low-temperature, or high-salt stress. *Plant Cell* 6: 251–264.

Zeevaart, J.A.D., Rock, C.D., Fantauzzo, F., Heath, T.G. and Gage, D.A. (1991) Metabolism of ABA and its physiological implications. *In:* W.J. Davies and H.G. Jones (eds): *Abscisic Acid: Physiology and Biochemistry,* Bios, Oxford, pp 39–52.

Signal Transduction in Plants
P. Aducci (ed.)
© 1997 Birkhäuser Verlag Basel/Switzerland

Auxin perception and signal transduction

M.A. Venis and R.M. Napier

Horticulture Research International, Wellesbourne, Warwick, CV35 9EF, UK

Introduction

A hormone receptor is a specific cellular recognition protein that binds the hormone and in consequence initiates a sequence of events (signal transduction) that culminates in a characteristic physiological or biochemical response. In the case of auxins there have been numerous reports of auxin-binding proteins (ABPs) over the last 25–30 years, but only one of these proteins has been studied in detail and has serious claims to be an auxin receptor. Hertel et al. (1972) found that auxin binding could be detected readily in preparations of maize (corn, *Zea mays* L.) microsomal membranes and soon several other laboratories were at work extending this initial characterisation. During the 1980s, this led to complete purification, cloning, and sequencing of what is now generally termed maize ABP1. Homologs have been found in many other species and during the last few years increasingly direct evidence for receptor function has appeared. This article will be largely confined to a review of our current understanding of ABP1 biochemistry and the evidence that this protein acts as an auxin receptor, together with the few clues available as to signaling intermediates that may be involved in transducing the hormonal signal following receptor activation. In addition, the nature of so-called ABPs detected by photoaffinity labelling with the 5-azido derivative of indole-3-acetic acid (IAA) will be discussed.

Biochemistry of ABP1

Developing from the original observations of Hertel et al. (1972), evidence for binding site heterogeneity was obtained, based on differences in affinity, specificity, and localization (sites 1 and 2, Batt et al., 1976; sites I, II, and III, Dohrmann et al., 1978). Site III was later shown to represent auxin accumulation in membrane vesicles (Hertel et al., 1983). Heterogeneity of binding sites was also observed by Jones et al. (1984a) and Shimomura et al. (1988) but not by Ray et al. (1977a) or Murphy (1980). It is generally agreed that the bulk of the binding activity (represented by site 1/I) is associated with endoplasmic reticulum (ER), but that auxin binding sites are located also on other membranes, variously suggested as plasma membrane (Batt and Venis, 1976), Golgi/-plasma membrane (Ray, 1977), or tonoplast (Dohrmann et al., 1978). As will be seen, there is strong evidence that a functional ABP population *is* present at the surface of the plasma membrane and is immunologically related to that in the ER. So far, there is no evidence that distinct subtypes of this ABP are present in different cellular membranes and because the purified protein resembles site 1/I in affinity (K_d around 0.1 μM for 1-naphthalene acetic acid, NAA) and predominant ER localization, it is now referred to as Zm-ERabp1 (Schwob et al., 1993) or, more briefly, ABP1.

Binding specificity for different auxins and analogs has been examined in greatest detail in total microsomal fractions, especially by Ray et al. (1977b), Dohrmann et al. (1978), and Murphy (1980). There is a reasonable correlation between physiological auxin activity and apparent K_d values, one of the major discrepancies being the relatively low binding affinity for the highly active 2,4-dichlorophenoxyacetic acid (2,4-D) and related analogs. The biological activity-affinity correlation was improved in the presence of a so-called supernatant factor (Ray et al., 1977a,b; Dohrmann et al., 1978), which was soon identified (Venis and Watson, 1978) as a mixture of benzoxazolinones, formed from the parent benzoxazinones during extraction. However, as the species distribution of this group of compounds is restricted to a few members of the Gramineae, they cannot have a general role in modulating auxin-ABP1 interaction.

Auxin binding activity in the membranes can be readily solubilized by detergent (Batt et al., 1976; Ray et al., 1977a; Cross et al., 1978), but the basis of most subsequent purification procedures has been a modified acetone powder method (Venis, 1977) that allows extraction without detergent. Initial purification by ion exchange and gel filtration (Venis, 1977) indicated an apparent native M_r of 40000 to 45000. The first extensive purification of ABP1 used a sequence of auxin-affinity and immuno-affinity columns (Löbler and Klämbt, 1985). The predominant polypeptide ran at an apparent M_r of 20000 on sodium dodecyl sulfate-polyacrylamide gel electrophoresis (SDS-PAGE), suggesting that ABP1 is a dimer. Subsequent purification protocols used either affinity chromatography (Shimomura et al., 1986; Radermacher and Klämbt, 1993), or else conventional chromatographic media in combination with native PAGE (Napier et al., 1988). Generally, a slightly smaller polypeptide (by ~1 kDa) copurifies with ABP1 (Shimomura et al., 1986, 1988; Napier et al., 1988). This appears to be a facile cleavage product of ABP1, lacking the C-terminus (Napier and Venis, 1990). Two minor isoforms, together representing <5% of the main ABP1 species can also be resolved (Hesse et al., 1989; Venis et al., 1992). Most laboratories find that ABP1 runs with an apparent M_r of 22000 on SDS-PAGE and hence it is usually

Table 1. Properties of Maize ABP1

Apparent native M_r	44 000
Apparent subunit M_r	22 000
Glycan M_r	2 000
Deduced sequence	163 residues
	+ 38 residue signal peptide
	single glycosylation site
	3 cysteines
	C-terminal KDEL (Lys-Asp-Glu-Leu)
Location	Endoplasmic reticulum
	Plasma membrane?
K_d (NAA)	0.1 to 0.2 µM (membrane)
	0.05 µM (purified)

referred to as 22 kDa ABP. It is still uncertain as to whether there is a single auxin-binding site per 22 kDa subunit (Hesse et al., 1989) or one binding site per dimer (Löbler and Klämbt, 1985; Shimomura et al., 1986).

In maize the various isoforms of ABP1 are encoded by a small gene family with five members identified so far (Lazarus et al., 1991; Yu and Lazarus, 1991; Schwob et al., 1993). The complete primary amino acid sequence of the major isoform/gene product ZmERabp1 was deduced by sequencing of cDNA clones (Inohara et al., 1989; Hesse et al., 1989; Tillman et al., 1989; Lazarus et al., 1991). The sequence indicates a 38-residue hydrophobic signal peptide, followed by the 163 residues of the mature (as extracted) protein. The mature protein is predominantly hydrophilic, with no obvious membrane-spanning domain and contains three cysteine residues. There is a single potential N-glycosylation site, consistent with a reduction in M_r of ~2000 following endoglycosidase H digestion (Löbler et al., 1987; Napier et al., 1988). The glycan is of the high mannose type (Hesse et al., 1989). The only other feature of particular note is a C-terminal KDEL tetrapeptide (lys-asp-glu-leu), a sequence shared by proteins that are actively retained within the lumen of the ER (Pelham, 1989). This feature is consistent, therefore, with the predominant ER localization deduced from earlier subcellular fractionation studies. The various properties of ABP1 are summarized in Table 1. There appears to be only a single *abp1* gene in *Arabidopsis* (Palme et al., 1992; Shimomura et al., 1993), while two genes have been described from tobacco (Shimomura, 1993) and a small gene family in strawberry (Lazarus et al., 1991; Lazarus and Macdonald, 1996).

The proteins encoded by the ER*abp* genes are highly conserved, the greatest variation occurring between signal (transit) peptide sequences. There is one stretch of complete sequence conservation over 10 residues, which is thought to embrace the auxin-binding site (Venis et al., 1992; discussed later) and several other stretches where conservation is almost complete or only conservative substitutions occur. The functional significance of these other conserved regions is not known. In addition, the C-terminal KDEL sequence for ER retention is always present. Dicotyledenous sequences contain additional potential N-glycosylation sites and in strawberry it appears that these are used (Lazarus and Macdonald, 1996).

Antibodies have proved to be of great value in structural and functional analysis of ABP1 and homologs. Polyclonal antisera to maize ABP1 have been produced in several laboratories (Löbler and Klämbt, 1985; Napier et al., 1988; Shimomura et al., 1988) and shown by Western blotting to cross-react in a range of species, including dicotyledonous species (Venis and Napier, 1990; Venis et al., 1992). Usually, the ABP1 homologs detected are the same subunit size as in maize (i.e., 22 kDa), but in some cases (e.g., barnyard grass, Venis and Napier, 1990) or mung bean (Napier and Venis, 1992) the apparent subunit size is slightly larger at 24 kDa. The difference, at least in the case of barnyard grass, is in the size of the polypeptide rather than the glycan. Next to maize, barnyard grass is the richest source of ABP1 and the relative abundance of immunoreactive polypeptides agrees well with relative auxin binding activities in microsomal preparations from the two species (Venis and Napier, 1990). Using an epitope mapping kit, three predominant linear epitopes in maize ABP1 were shown to be recognized by anti-ABP1 sera from several different laboratories (Napier and Venis, 1992). These epitopes are clustered around, but do not include, the glycosylation site and appear to be regions that are exposed on the surface of the protein. Two of these three epitopes are conserved in ABP1 homologs from mung bean and barnyard grass.

A set of five monoclonal antibodies against maize ABP1 has been raised, designated MAC 256

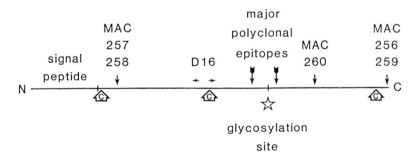

Figure 1. Main structural features of ABP1 and the regions recognised by antibodies. The positions of the three cysteine residues (C) are indicated by the bold arrows. The region designated D16 indicates the putative auxin-binding site, being the part of the sequence against which antibodies (named D16) showing auxin agonist activity were raised (Venis et al., 1992).

through MAC 260 (Napier et al., 1988). The epitopes recognized by these antibodies were assigned by fragmentation studies (Napier and Venis, 1990) in conjunction with epitope mapping (Napier and Venis, 1992). Of particular interest are MAC 256 and MAC 259, which are specific for the C-terminal region, especially the ER retention sequence KDEL. In consequence, these antibodies recognize ER-resident proteins in animal cells and are excellent markers for animal cell ER (Napier et al., 1992). Plants also use HDEL for ER retention and ABP1 appears to be the major KDEL protein in maize ER (Napier et al., 1992). The major structural features and epitopes of ABP1 are summarized in Figure 1.

Using a sandwich ELISA (enzyme-linked immunosorbent assay), it was found that binding of these two monoclonal antibodies – MAC 256 and MAC 259 – to native ABP1 was reduced by auxins and analogs in a concentration-dependent manner (Napier and Venis, 1990). There was an excellent correlation between this activity and the physiological activity of the wide range of compounds tested. Indeed, the structure-activity correlation was better than that obtained from *in vitro* assays of labelled NAA binding to microsomal or solubilized ABP1, for example, phenoxyacetic acids such as 2,4-D were about as active in the ELISA as NAA. It appears that the presentation of ABP1 in the ELISA may reflect more accurately the *in vivo* conformation of the protein. The auxins and monoclonal antibodies were not thought to be competing for the same binding site, and the reduction in antibody binding was interpreted as an auxin-induced conformational change that leads to masking of the epitope to which the antibody binds. Because this epitope is at the C-terminus, the KDEL region appears to be conformationally active, and this, as discussed later, may have important mechanistic implications. Evidence for auxin-induced conformational change was also obtained by Shimomura et al. (1986) from circular dichroism spectroscopy.

Maize ABP1 has been expressed in the baculovirus system (Macdonald et al., 1994), yielding a product which is glycosylated, binds auxin, and is correctly targeted to the lumen of the ER. In addition strawberry ABP1 has now been expressed in baculovirus (Lazarus and Macdonald, 1996), opening up possibilities both for crystallisation and biochemical characterisation of monocot and dicot ABP1 s.

Use of ^{3}H-5-azido-IAA

Maize

This analog was synthesized by Melhado et al. (1981) as a potential auxin affinity label. It was first used by Jones et al. (1984b) with maize microsomes and later with partly purified ABP1

M. A. Venis and R. M. Napier

preparations, resulting in radiolabelling of 22 kDa ABP1, together with a 24 kDa polypeptide (Jones and Venis, 1989). The latter polypeptide has recently been identified as a manganese superoxide dismutase (Feldwisch et al., 1995). Photolabelling of pure maize ABP1 followed by tryptic digestion resulted in labelling predominantly of a single polypeptide (Brown and Jones, 1994). A specific aspartate residue (asp[134]) was proposed as the target of azido-IAA, close to a proposed hydrophobic platform (Edgerton et al., 1994) of the auxin-binding site. On the other hand, NAA blocked photolabelling far less effectively than IAA (Brown and Jones, 1994). Since reversible binding studies have always shown that the affinity of ABP1 for NAA is about ten times greater than for IAA (e.g., Löbler and Klämbt, 1985a, and references therein), it is possible that 5-azido-IAA does not in fact target the high affinity auxin-binding site of ABP1 under these conditions.

In the above cases, photolysis was carried out at 0–4°C, but somewhat different results have been obtained after labelling at –196°C (liquid nitrogen). Thus, in partially purified ABP1 preparations labelling of the 22 kDa subunit was efficiently competed by NAA (Campos et al., 1991) in contrast to labelling at 0°C, though the NAA concentration (100 µM) was ten times higher than that used by Brown and Jones (1994). A 60 kDa polypeptide was more heavily labelled in the same preparations, but NAA competed relatively poorly. Subsequent labelling studies with microsomal and soluble fractions showed that the 60 kDa polypeptide (p 60) was predominantly cytosolic. Sequencing indicated that the microsomal form (pm 60) was very similar to the cytosolic form (p 60 s), but with an extra five residues at the N-terminus (Moore et al., 1992). The protein was purified from the microsomal fraction and was found to show sequence homologies with glucosidases (Campos et al., 1992). One peptide sequence was identical to a peptide in the *rol C* gene product, a protein that is able to hydrolyze a variety of cytokinin O- and N-glucosides (Estruch et al., 1991). The p 60 protein showed β-glucosidase activity against *p*-nitro-phenyl-β-D-glucoside and against indican (indoxyl-β-D-glucoside), but did not hydrolyze the *myo*-inositol or aspartate conjugates of IAA, nor rutin, a quercetin glucoside (Campos et al., 1992). Apart from azido-IAA itself and 2-NAA, auxins (2,4-D, IAA, NAA) even at 1 mM reduced ^{3}H-azido-IAA incorporation into p 60 by only 44 to 63%. The protein did not show equilibrium binding of IAA and very high auxin concentrations were needed to inhibit its glucosidase activity (60% inhibition at 2 mM NAA). Campos et al. (1992) first suggested that p 60 might have a role in hydrolysis of IAA conjugates and thus influence cellular free IAA concentrations, but there is little support for such a role from the substrate specificity described. Subsequently, the recombinant p 60 protein was shown to hydrolyze zeatin-*O*-glucosides and kinetin-N3-glucoside, but not cytokinin-N7 or N9-glucosides (Brzobohaty et al., 1993). While p 60 may, as the authors suggested, have a function in releasing active cytokinins, the significance of p 60 labelling by azido-IAA is still unclear.

Similar azido-IAA photolabelling (at $-196°C$) of maize plasma membrane preparations by the same research group (Feldwisch et al., 1992) showed that in addition to predominant labelling of a 60 kDa polypeptide, a 23 kDa band on SDS-PAGE (pm 23) was faintly labelled. In microsomal fractions a radioactive polypeptide at 24 kDa (pm 24) was also detected. Both polypeptides were extracted from microsomal membranes and purified to homogeneity. Photolabelling of pm 23 in maize plasma membrane fractions was efficiently competed by auxins at 1 mM (IAA, NAA, 2,4-D), but not by 2-NAA (Feldwisch et al., 1992). In addition, pm 23 labelling was competed by the auxin transport inhibitors 2,3,5-triiodobenzoic acid (TIBA) and N-1-naphthylphthalamic acid (NPA) (Feldwisch et al., 1992) and what appears to be the same 23 kDa polypeptide in maize plasma membranes can also be photolabelled with ^{3}H-azido-NPA (Zettl et al., 1992). Thus, the pm 23 polypeptide seems to contain binding sites for both IAA and NPA and may be a component of the auxin efflux carrier. In passing, it should be noted that azido-NPA labels predominantly a 60 kDa polypeptide, followed by a 24 kDa band; pm 23 is labelled the least. The larger azido-NPA-labelled polypeptides were designated pm 60 and pm 24, although it is not clear whether these correspond to the equivalent polypeptides labelled with azido-IAA (Feldwisch et al., 1992). Unlike pm 23, labelling of pm 24 and pm 60 by azido-NPA was not affected by NPA, TIBA, or IAA and was regarded as nonspecific (Zettl et al., 1992).

Zucchini and tomato

With plasma membrane preparations of zucchini hypocotyls, Hicks et al. (1989a) found that at $-196°C$ azido-IAA labelling was confined to a pair of low abundance 40 and 42 kDa polypeptides that behaved as integral membrane proteins. IAA and 2-NAA largely blocked labelling, while 2,4-D and NAA (1-NAA) were less effective. This pattern of competition (especially the behaviour of the NAA isomers) is distinct from the specificity of ABP1 but is akin to that expected of the auxin uptake carrier.

Similar photolabelling experiments (at $-196°C$) also identified a 40 and 42 kDa doublet in microsomal membranes of tomato hypocotyls and roots (Hicks et al., 1989b). In the auxin-insensitive *diageotropica* mutant, labelling of hypocotyl preparations was very much reduced, compared with wild type, although in roots it was about the same. It was thought that the difference in labelling might be related to impaired auxin perception or transport underlying the mutant phenotype. Microsomal membrane proteins of several other species, including maize, were also photolabelled by azido-IAA in the 40 to 42 kDa range (Lomax and Hicks, 1992). This paper also reported that anti-ABP1 sera did not cross-react with the photolabelled polypeptides of zucchini, again indicating that they are distinct from ABP1. In discussing the possible role of the 40 and

42 kDa polypeptides, Lomax and Hicks (1992) concluded that they are more likely to function in auxin uptake than as auxin receptors. Subsequent analysis of labelled zucchini preparations on 2-D gels revealed several isoforms of the 40 and 42 kDa polypeptides, the major isoforms having pI values of 8.2 and 7.2, respectively (Hicks et al., 1993). This is a further distinguishing feature from ABP1, which is much more acidic (Jones and Venis, 1989). From gel filtration and partitioning properties Hicks et al. (1993) suggested that the 40 and 42 kDa polypeptides may be related subunits of a multimeric channel protein involved in auxin uptake.

Hyoscyamus and Arabidopsis

When azido-IAA was used with soluble fractions of *Hyoscyamus* suspension culture cells, polypeptides of 31 and 24 kDa were photolabelled at 0°C, while at -196°C a 25 kDa polypeptide was labelled (Macdonald et al., 1991). None of these polypeptides was recognized by antibodies to maize ABP1. Labelling of the 31 and 24 kDa bands was efficiently competed by IAA ($K_i \sim 3$ µM) but not by other auxins (NAA, 2,4-D) or by 2-NAA. It was found that these two polypeptides were immunologically related to a tobacco β-1,3-glucanase and that the purified tobacco enzyme could also be photolabelled by azido-IAA. IAA also reduced labelling of the 25 kDa polypeptide, although far less effectively ($K_i > 100$ µM), while NAA, 2-NAA and, to a lesser extent, 2,4-D competed to some degree at 0.2 mM. Subsequently (Bilang et al., 1993), this polypeptide was purified and shown to be glutathione *S*-transferase (GST). Purified GST showed the same labelling characteristics towards azido-IAA. It was not possible to detect binding of ^{3}H-IAA to the 25 kDa GST polypeptide by equilibrium dialysis. More recently, a 24 kDa polypeptide labelled by azido-IAA in *Arabidopsis* was also identified as a GST (Zettl et al., 1994), although in this case the enzyme was membrane bound.

Conclusions

It is clear that results from photolabelling with tritiated 5-azido-IAA (summarized in Tab. 2) need to be interpreted with care. In the absence of other information, the covalent attachment of azido-IAA to a polypeptide does not necessarily identify the target as a binding site specific for auxin, and hence the polypeptide may have no role in auxin transport or perception. It is not clear why different laboratories using similar membrane fractions from the same species (maize) under similar conditions report labelling of 23, 24, 60 kDa polypeptides on the one hand (Feldwisch et al., 1992) and 40 and 42 kDa polypeptides on the other (Lomax and Hicks, 1992). Other than

Table 2. Polypeptides photolabelled by ^{3}H-5-azido-IAA

Cell fraction	Labelling		Ref.
	Size, kDa	Identity	
Microsomes; plasma membrane	22	ABP1	Jones and Venis, 1989; Campos et al., 1991; Brown and Jones, 1994
Microsomes; plasma membrane	40, 42	Auxin uptake carrier?	Hicks et al., 1989a, b, 1993; Lomax and Hicks, 1992
Plasma membrane	23	Auxin efflux carrier?	Feldwisch et al., 1992
Microsomes	23	Mn-superoxide dismutase	Feldwisch et al., 1995
Microsomes	24	?	Feldwisch et al., 1992
Soluble; microsomes	60	Glucohydrolase	Campos et al., 1992; Moore et al., 1992; Brzobohaty et al., 1993
Soluble	24, 31	Glucanase	Macdonald et al., 1991
Soluble	25	Glutathione S-transferase	Bilang et al., 1993
Plasma membrane	24	Glutathione S-transferase	Zettl et al., 1994

ABP1, in the five cases where the labelled polypeptides have been characterized (immunologically or by sequencing) their identities – glucohydrolase, superoxide dismutase, glucanase and GST (two species) – suggest they are unlikely to have any direct role in auxin perception. It is worth mentioning here that a 65 kDa polypeptide identified as a putative ABP by an anti-idiotypic approach (Prasad and Jones, 1991) also shows GST homology (Jones and Prasad, 1994).

The indirect evidence available does suggest that the 40 and 42 kDa doublet and the 23 kDa membrane polypeptide are likely to be connected with auxin uptake and efflux, respectively. Such proteins are undoubtedly of profound significance for auxin action and the azido-IAA tag may well play an important part in their further characterization. The compound has also proven of value in tracing the path of auxin transport through tissues (Jones, 1990). Nevertheless, as a probe for proteins involved in auxin perception (i.e., as a receptor probe), the usefulness of azido-IAA remains to be established.

Is ABP1 an auxin receptor?

Until a few years ago, the view that ABP1 might be an auxin receptor was based largely on indirect physiological correlates between binding activity or ABP1 abundance and auxin responsiveness. Latterly, however, electrophysiological studies have provided more direct and persuasive evidence, using initially the characteristic auxin-induced hyperpolarization of the membrane potential of tobacco protoplasts (Ephritikhine et al., 1987). First it was shown that the auxin-induced hyperpolarization response was blocked by monospecific antibodies to maize ABP1 and also by anti-H$^+$ATPase (Barbier-Brygoo et al., 1989). This suggested that a tobacco ABP1, immunologically related to maize ABP1, is located on the exterior face of the plasma membrane, is coupled to a H$^+$ ATPase and is essential for the auxin-induced response. Subsequent sequencing has confirmed extensive homologies between maize and tobacco ABP1 s (Shimomura, 1993), a tobacco ABP1 homolog has been detected by immunoblotting (Venis et al., 1992) and work with impermeant auxin analogs provided further evidence for a site of auxin perception at the cell surface (Venis et al., 1990). Additional support for a role of ABP1 in mediating auxin perception at the plasma membrane arose from experiments showing that the auxin sensitivity of the protoplast hyperpolarization response could be manipulated experimentally through several orders of magnitude. Thus, in the presence of a low concentration (0.4 nM) of anti-ABP1 IgG, the NAA concentration needed to give an optimal response was increased 10-fold; conversely, the optimum concentration was reduced 1000-fold when purified maize ABP1 was added to the protoplast medium at 1 pM (Barbier-Brygoo et al., 1991).

The next significant development was the generation of anti-ABP1 antibodies with auxin agonist activity (Venis et al., 1992). From early experiments with group-modifying reagents, provisional assignments of six of the amino acids likely to be present at the auxin binding site of ABP1 had been made. Inspection of the deduced amino acid sequence of ABP1 showed that five of these were clustered in a single hexapeptide. Antisera (e.g., D16) raised against a synthetic oligopeptide embracing this region recognized all maize ABP1 isoforms as well as ABP1 homologs in several other species, indicating that the region selected was likely to be highly conserved. More significantly, the anti-peptide antibodies were able to mimic precisely the activity of auxin in hyperpolarizing the transmembrane potential of tobacco protoplasts (Venis et al., 1992). This auxin agonist activity strongly suggested that the selected peptide lies in the auxin-binding domain of an auxin receptor. The likely importance of this region is reinforced by subsequent data showing that there is almost complete sequence conservation between species (Fig. 2) and that it contains the longest stretch of wholly conserved sequence in ABP1.

The conclusions reached on the basis of the tobacco protoplast hyperpolarization response have been fully supported by subsequent independent measurements of membrane current in maize protoplasts. Using the patch-clamp technique in the whole-cell configuration, an auxin-induced increase in outwardly directed current of positive charge was detected under conditions consistent with stimulation of the plasma membrane H$^+$-ATPase (Rück et al., 1993). This auxin-induced current was blocked by anti-ABP1 antibodies, while the anti-peptide antibodies raised against the putative auxin-binding domain showed auxin agonist activity, stimulating the membrane current in the absence of auxin. Thus, both agonist and antagonist activities of ABP1-related antibodies on

Peptide	**RTPIHRHSCEEVFT**		
Zea (*Zmabp1*)	**PGQ**	**RTPIHRHSCEEVFT**	**VLKG**
Nicotiana	**PGS**	**RTPIHRHSCEEIFI**	**VLKG**
Arabidopsis	**PGS**	**ETPIHRHSCEEVFV**	**VLKG**
Fragaria	**PGS**	**GTPIHRHSCEEVFV**	**VLKG**

Figure 2. Sequence of peptide synthesized for antibody production compared with ABP1 sequences from four different species.

an auxin-dependent physiological response have been demonstrated in two different systems, one homologous (maize protoplasts with anti-maize ABP1 antibodies, Rück et al., 1993) and one heterologous (tobacco protoplasts, maize antibodies).

More recently, direct chemical evidence supporting the physiological relevance of ABP1 has been obtained. Six peptides corresponding to predicted surface domains of ABP1 were synthesized and tested for effects on ion channel activities in the plasma membrane of *Vicia* guard cells (Thiel et al., 1993). Of these, only one – the 13 residue C-terminus of ABP1 – had biological activity, the dominant effect being inactivation of the inward-rectifying K^+ channel current, with a half-maximal effect at 16 μM peptide. The peptide also caused a rapid and reversible alkalinization of cytoplasmic pH, averaging 0.4 pH units. When shifts in internal pH were buffered by butyrate, the peptide effect on K^+ current was also blocked. The changes in K^+ current evoked by peptide were similar to those seen at 30 to 100 μM auxin, this response also being antagonized in the presence of butyrate (Blatt and Thiel, 1994). The peptide has since been found to inhibit auxin-induced elongation growth of maize coleoptiles in a non-competitive manner, with an I_{50} of ca. 10 μM (M.R. Blatt, personal communication and poster presentation at Plant Membrane Biology Workshop, Regensburg, Germany, 1995). While the relationship of the peptide effect to normal auxin signalling cannot be interpreted precisely at this stage, these results are important in linking a defined sequence of ABP1 to auxin-related responses.

How does ABP1 get to the plasma membrane?

The electrophysiological data lead to the conclusion that there is a functional pool of ABP1 on the outside surface of the plasma membrane yet this protein is actively targeted to the ER. Every sequenced member of the ABP1 gene family in all species contains both ER transit and retention signals, consistent with all reports on subcellular compartmentation of maize ABP1, showing that at least 90% of the protein comigrates with ER marker activities (Ray, 1977; Shimomura et al., 1988; Jones et al., 1989; Napier et al., 1992), although some ABP1 can be identified with denser fractions, possibly plasma membrane. The small pool size at the plasma membrane relative to the much larger ER pool has made it extremely difficult to demonstrate passage of ABP1 to the cell surface.

In order to reach the plasma membrane, the commitment to targeting conferred by KDEL has to be overcome. Our earlier observations, which suggested that ligand binding induced a confirmational change masking KDEL (Napier and Venis, 1990), presented a mechanism that can explain ABP1 escape and such a model has been elaborated by Cross (1991). Once the KDEL retrieval system has been bypassed, ABP1 would continue to the cell surface through the constitutive secretory pathway. No *in vivo* evidence to support the model that release is triggered by auxin is available yet. One report does claim to show that ABP1 is secreted along the constitutive pathway (Jones and Herman, 1993). In this paper, ultrathin sections of maize cells were examined after immunogold labelling using affinity-purified ABP1 antibodies. Gold particles were found throughout the cell wall and were concentrated along the plasma membrane. Gold was also seen decorating the Golgi and some was found in the cytoplasm associated with the ER. The authors argued that the heavy cell wall and plasma membrane staining showed that ABP1 is secreted and that the Golgi staining shows that the constitutive pathway is used, but the immunogold decoration described is exactly the opposite of that expected from all the cell fractionation studies, in particular the heavy plasma membrane decoration. Our own data with ABP1 monoclonal antibodies for immunocytology using both fluorescence and electron microscopy support the primarily intracellular (ER) localization of ABP1 (Napier et al., 1992; Henderson, Hawes, Napier and Venis, unpublished). In addition to microscopy, cells and culture media from maize suspension cells were analyzed by SDS-PAGE and immunoblotting (Jones and Herman, 1993). A 24 kDa peptide was detected in both cells and media and its abundance in the medium increased with days of culture. Its appearance in the medium was blocked by Brefeldin A (which inhibits secretion) when the cells were cultured for 72 hours in the drug and so the 24 kDa polypeptide certainly appears to be a secretory protein. However, there is an important anomaly in the data, in that the 24 kDa polypeptide detected is significantly larger than authentic 22 kDa maize ABP1, a mixture of the two running as a doublet, not a single band (Jones and Herman, 1993).

Labelling of maize coleoptile ABP1 *in vivo* (Oliver et al., 1995) showed that the protein is very stable (half-life >24 hours), that its synthesis and turnover were unaffected by auxin and that no auxin-induced change in labelled ABP1 could be detected in plasma membrane-rich fractions, even after prolonged incubations. These data probably preclude large-scale efflux of ABP1 to the plasma membrane, whether constitutive or assisted by auxin-induced conformational change (Napier and Venis, 1990), but cannot rule out the possibility of a very low level of efflux, below the detectability of the protocols used.

Hence, at present the biochemical case for passage of ABP1 from ER to plasma membrane must be regarded as non-proven. Despite this uncertainty, it has in fact proved possible to detect the small population of ABP1 at the external surface of maize coleoptile protoplasts by the technique of SEIG-EPOM (*silver-enhanced immunogold* viewed by *epipolarization microscopy*) (Fig. 3). While the published number of specific antibody-binding sites was ca. 800 (Diekmann et al., 1995) a magnification error means the true number is four to five-fold higher (S. Rinke and D.G. Robinson, personal communication), which would represent roughly 2% of total cellular ABP1. Note that the sensitivity of the technique, which views the whole surface, is inherently much greater than immunogold decoration of sections. A noteworthy finding in the same report was that preincubation of the protoplasts with auxin induced a dramatic clustering of the signal. This response was time- and temperature-dependent and occurred only with physiologically active auxins (Diekmann et al., 1995). While the precise significance of this response awaits further elucidation, ligand-induced receptor clustering is a feature of the action of many animal hormones and is also time- and temperature-dependent (Metzger, 1992). Since ABP1 clustering takes 30–60 min at 25°, this response is clearly not required for the electrophysiological H^+ translocation responses discussed earlier, which are complete in 1–2 min (Barbier-Brygoo et al., 1989; Rück et al., 1993). However, other auxin-responses might be dependent on ABP1 clustering.

What happens next?

Inasmuch as ABP1 behaves as a soluble protein and has no obvious transmembrane domain, it has been suggested that it interacts through a "docking protein" in the plasma membrane with signaling elements (Klämbt, 1990; Barbier-Brygoo et al., 1991). The biological activity of the C-terminal ABP1 peptide (Thiel et al., 1993) suggests that this domain might be involved in such an interaction and it is of considerable interest that it is also the C-terminal region of ABP1 that appears to undergo an auxin-induced conformational change, as discussed earlier (Napier and Venis, 1990). Thus, this region has the right credentials for linking auxin-ABP1 interaction to intracellular signaling cascades. The work with the C-terminal peptide (Thiel et al., 1993) together with other studies (discussed in Napier and Venis, 1995) has implicated modulation of intracellular pH as part of the auxin signaling cascade. Other possible signaling intermediates will now be discussed briefly.

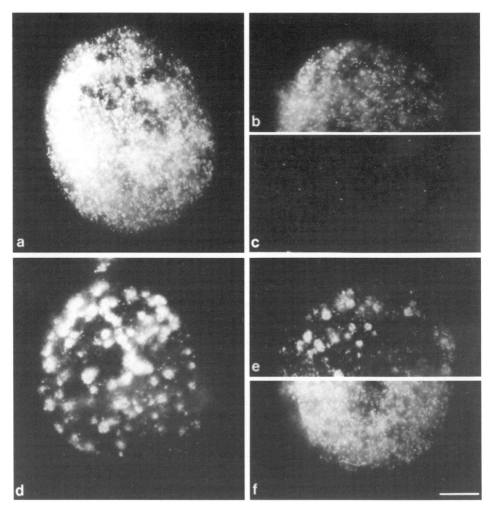

Figure 3. (a) Maize coleoptile protoplast treated with anti-ABP1 antibodies and visualised by SEIG-EPOM. (b) As a, but with D16 antibodies. (c) As b but with D16 + 10 μM IAA. (d) As a, but protoplast treated with 10 μM IAA (1 hour, 25°C) before incubation in anti-ABP1. (e) As d, but with 1-NAA instead of IAA. (f) As d, but with 2-NAA (adapted from Diekmann et al., 1995).

Auxin-induced changes in intracellular Ca^{2+} concentration appear to be associated with pH changes (Felle, 1988; Gehring et al., 1990; Irving et al., 1992). In each case cytoplasmic acidification was correlated with increased cytoplasmic $[Ca^{2+}]$ and where pH was shown to oscillate, Ca^{2+} changed in concert (Felle, 1988). Most of the components necessary for a phosphoinositide signaling pathway to induce intracellular Ca^{2+} release have been shown to be present in plants as in mammals. Auxin was reported to induce a rapid, transient rise in phospholipase C activity, leading to increases in inositol (1,4,5) trisphosphate (IP_3) and inositol bisphosphate (Ettlinger and Lehle, 1988; Zbell and Walter-Back, 1988), but there has been no development or confirmation of this work.

There is little doubt that phosphoinositides can induce Ca^{2+} release from plant cell vesicles (Drobak and Ferguson, 1985; Schumaker and Sze, 1987; Gilroy et al., 1990) and an IP_3-gated calcium channel in red beet vacuoles has been described (Alexandre et al., 1990) as well as an IP_3 binding protein in mung bean (Biswas et al., 1995). However, the putative IP_3 pool size in plants is very small and turnover rapid (Drobak et al., 1991) and, without further characterization, the involvement of the phosphoinositide pathway in auxin signal transduction remains equivocal.

There is growing evidence for the involvement of the phospholipase A_2 pathway in auxin signal transduction. Auxin has been shown to raise the activity of phospholipase A_2 by up to 70%, the rise being apparent within minutes and showing good specificity for active auxins (Scherer and André, 1989, 1993; André and Scherer, 1991). Further, a membrane-associated protein kinase was shown to be activated by lysolipids and microsomal H^+-ATPase activity raised, possibly *via* phosphorylation by the kinase (Martiny-Baron and Scherer, 1989). Phospholipase A_2 activation has been linked to auxin perception by ABP1, antibodies to ABP1 inhibiting the rise in activity induced by auxin (Scherer and André, 1993) and further characterization of this pathway is awaited with interest.

In animal systems, signaling by the seven transmembrane span (7TMS) receptor superfamily to, for example, phosphoinositide metabolism is coupled by heterotrimeric GTP-binding proteins or G-proteins (Hepler and Gilman, 1992). Such proteins are also present in plants and have been implicated in a range of growth and developmental processes (Terryn et al., 1993). However, there is so far only one, unconfirmed report linking GTP binding activity with auxin signal tranduction (Zaina et al., 1990). Nevertheless, two pieces of evidence point to the existence of plant homologues of G-protein-coupled 7TMS receptors. First, affinity chromatography using a C-terminal (putative receptor-interacting) peptide of *Arabidopsis* G-protein type α1 yielded a 37 kDa polypeptide from maize microsomal extracts (Wise et al., 1994), though this remains uncharacterized, apart from N-terminal sequence data. Second, the peptide mastoparan, which mimics the G-protein interacting domain of 7TMS receptors, was found to stimulate binding of GTP-γ-S to pea and maize membranes (White et al., 1993) and also to stimulate both growth and phospholipase A_2 activity in sections and suspension cells (Scherer, 1992, 1995).

An important recent development has been the cloning of the *AUX1* gene of *Arabidopsis*, which appears to be involved in regulating auxin-dependent root growth and in hormone control of gravitropism. Sequence analysis of *AUX1*, suggests that the gene product is a member of a family of amino acid permeases. It is proposed that IAA is a likely substrate and hence that AUX1 is involved in auxin transport (Bennett et al., 1996), though this remains to be established. This does not exclude the possibility that AUX1 may also represent the hypothetical docking protein for ABP1.

Conclusions

For rapid, H^+-extrusion responses there is convincing evidence that ABP1 on the plasma membrane is the site of auxin perception (but see Hertel, 1995 and Venis, 1995 for discussion of a contrary view). For all other responses, e.g., gene expression, no such connection to any auxin-binding protein can yet be made. Even for H^+-extrusion the signal transduction events between perception, conformational change and elevation of H^+-ATPase activity are uncertain. A major challenge ahead is to identify components of one signal transduction pathway so that intermediates of this pathway can be used to test for the specificity or diversity of a defined auxin binding event and components of the pathway.

Acknowledgements
Work from the authors' laboratory was supported by the Biotechnology and Biological Sciences Research Council and by the BAP, BRIDGE and BIOTECH programs of the European Economic Communities. We thank Drs Heather Macdonald and Colin Lazarus for supplying the strawberry ABP1 sequence prior to publication.

References

Alexandre, J., Lassalles, J.P. and Kado, R.T. (1990) Opening of Ca^{2+} channels in isolated red beet vacuole membrane by inositol (1,4,5)trisphosphate. *Nature* 343: 567–570.
André, B. and Scherer, G.F.E. (1991) Stimulation by auxin of phospholipase A in membrane vesicles from an auxin-sensitive tissue is mediated by an auxin receptor. *Planta* 185: 209–214.
Barbier-Brygoo, H., Ephritikhine, G., Klämbt, D., Ghislain, M. and Guern, J. (1989) Functional evidence for an auxin receptor at the plasmalemma of tobacco mesophyll protoplasts. *Proc. Natl. Acad. Sci. USA* 86: 891–895.
Barbier-Brygoo, H., Ephritikhine, G., Klämbt, D., Maurel, C., Palme, K., Schell, J. and Guern, J. (1991) Perception of the auxin signal at the plasma membrane of tobacco mesophyll protoplasts. *Plant J.* 1: 83–94.
Batt, S. and Venis, M.A. (1976) Separation and localization of two classes of auxin binding sites in corn coleoptile membranes. *Planta* 130: 15–21.
Batt, S., Wilkins, M.B. and Venis, M.A. (1976) Auxin binding to corn coleoptile membranes: kinetics and specificity. *Planta* 130: 7–13.
Bennett, M.J., Marchant, A., Green, H.G., May, S.T., Ward, S.P., Millner, P.A., Walker, A.R., Burkhard, S. and Feldman, K.A. (1996) *Arabidopsis AUX1* gene: a permease-like regulator of root gravitropism. *Science*; *in press*.
Bilang, J., Macdonald, H., King, P.J. and Sturm, A. (1993) A soluble auxin-binding protein from *Hyoscyamus muticus* is a glutathione *S*-transferase. *Plant Physiol.* 102: 29–34.
Biswas, S., Dalal, B., Sen, M. and Biswas, B.B. (1995) Receptor for *myo*-inositol trisphosphate from the microsomal fraction of *Vigna radiata*. *Biochem. J.* 306: 631–636.
Blatt, M. and Thiel, G. (1994) K^+ channels of stomatal guard cells: bimodal control of the K^+ inward-rectifier evoked by auxin. *Plant J.* 5: 55–68.

Brown, J.C. and Jones, A.M. (1994) Mapping the auxin-binding site of auxin binding protein 1. *J. Biol. Chem.* 269: 21136–21140.

Brzobohaty, B., Moore, I., Kristoffersen, P., Bako, L., Campos, N., Schell, J. and Palme, K. (1993) Release of active cytokinin by a β-glucosidase localized to the maize root meristem. *Science* 262: 1051–1054.

Campos, N., Feldwisch, J., Zettl, R., Boland, W., Schell, J. and Palme, K. (1991) Identification of auxin-binding proteins using an improved assay for photoaffinity labeling with 5-N₃[7-³H]-indole-3-acetic acid. *Technique* 3: 69–75.

Campos, N., Bako, L., Feldwisch, J., Schell, J. and Palme, K. (1992) A protein from maize labeled with azido-IAA has novel β-glucosidase activity. *Plant J.* 2: 675–684.

Cross, J.W. (1991) Cycling of auxin-binding protein through the plant cell: pathways in auxin signal transduction. *New Biol.* 3: 813–819.

Cross, J.W., Briggs, W.R., Dohrmann, U.C. and Ray, P.M. (1978) Auxin receptors of maize coleoptile membranes do not have ATPase activity. *Plant Physiol.* 61: 581–584.

Diekmann, W., Venis, M.A. and Robinson, D.G. (1995) Auxins induce clustering of the auxin-binding protein at the surface of maize coleoptile protoplasts. *Proc. Natl. Acad. Sci. USA* 92: 3425–3429.

Dohrmann, U., Hertel, R. and Kowalik, H. (1978) Properties of auxin binding sites in different subcellular fractions from maize coleoptiles. *Planta* 140: 97–106.

Drobak, B.K. and Ferguson, I.B. (1985) Release of Ca²⁺ from plant hypocotyl microsomes by inositol 1,4,5-trisphosphate. *Biochem. Biophys. Res. Comm.* 130: 1241–1246.

Drobak, B.K., Watkins, P.A.C., Chattaway, J.A., Roberts, K. and Dawson, A.P. (1991) Metabolism of inositol (1,4,5-) trisphosphate by a soluble enzyme fraction from pea roots. *Plant Physiol.* 95: 412–419.

Edgerton, M.D., Tropsha, A. and Jones, A.M. (1994) Modelling the auxin-binding site of auxin-binding protein of maize. *Phytochemistry* 35: 1111–1123.

Ephritikhine, G., Barbier-Brygoo, H., Muller, J.F. and Guern, J. (1987) Auxin effect on the transmembrane potential difference of wild-type and mutant tobacco protoplasts exhibiting a differential sensitivity of auxin. *Plant Physiol.* 84: 801–804.

Estruch, J.J., Chriqui, D., Grossmann, K., Schell, J. and Spena, A. (1991) The plant oncogene *rolC* is responsible for the release of cytokinins from glucoside conjugates. *EMBO J.* 10: 2889–2895.

Ettlinger, C. and Lehle, L. (1988) Auxin induces rapid changes in phosphatidylinositol metabolites. *Nature* 331: 176–178.

Feldwisch, J., Zettl, R., Hesse, F., Schell, J. and Palme, K. (1992) An auxin-binding protein is localized to the plasma membrane of maize coleoptile cells: identification by photoaffinity labeling and purification of a 23-kDa polypeptide. *Proc. Natl. Acad. Sci. USA* 89: 475–479.

Feldwisch, J., Zettl, R., Campos, N. and Palme, K. (1995) Identification of a 23 kDa protein from maize photoaffinity labeled with azido-IAA. *Biochem. J.* 305: 853–857.

Felle, H. (1988) Auxin causes oscillations of cytosolic free calcium and pH in *Zea mays* coleoptiles. *Planta* 174: 495–499.

Gehring, C.A., Irving, H. and Parish, R.W. (1990) Effects of auxin and abscisic acid on cytosolic calcium and pH in plant cells. *Proc. Natl. Acad. Sci. USA* 87: 9645–9649.

Gilroy, S., Read, N.D. and Trewavas, A.J. (1990) Elevation of cytoplasmic calcium by caged calcium or caged inositol trisphosphate initiates stomatal closure. *Nature* 346: 769–771.

Hepler, J.R. and Gilman, A.G. (1992) G-proteins. *Trends Biochem. Sci.* 17: 383–387.

Hertel, R. (1995) Auxin binding protein 1 is a red herring. *J. Exp. Bot.* 46: 461–462.

Hertel, R., Thomson, K.-St. and Russo, V.E.A. (1972) *In vitro* auxin binding to particulate cell fractions from corn coleoptiles. *Planta* 107: 325–340.

Hertel, R., Lomax, T.L. and Briggs, W.R. (1983) Auxin transport in membrane vesicles from *Cucurbita pepo* L. *Planta* 157: 193–201.

Hesse, T., Feldwisch, J., Balschusemann, D., Bauw, G., Puype, M., Vandekeckhove, J., Löbler, M., Klämbt, D., Schell, J. and Palme, K. (1989) Molecular cloning and structural analysis of a gene from *Zea mays* (L.) coding for a putative receptor for the plant hormone auxin. *EMBO J.* 8: 2453–2461.

Hicks, G.R., Rayle, D.L., Jones, A.M. and Lomax, T.L. (1989a) Specific photoaffinity labeling of two plasma membrane polypeptides with an azido auxin. *Proc. Natl. Acad. Sci. USA* 86: 4948–4952.

Hicks, G.R., Rayle, D.L., Jones, A.M. and Lomax, T.L. (1989b) The *diageotropica* mutant of tomato lacks a high specific activity auxin binding site. *Science* 245: 52–54.

Hicks, G.R., Rice, M.S. and Lomax, T.L. (1993) Characterization of auxin-binding proteins from zucchini plasma membrane. *Planta* 189: 83–90.

Inohara, N., Shimomura, S., Fukui, T. and Futai, M. (1989) Auxin-binding protein located in the endoplasmic reticulum of maize shoots: molecular cloning and complete primary structure. *Proc. Natl. Acad. Sci. USA* 83: 3564–3568.

Irving, H.R., Gehring, C.A. and Parish, R.W. (1992) Changes in cytosolic pH and calcium of guard cells precede stomatal movements. *Proc. Natl. Acad. Sci. USA* 89: 1790–1794.

Jones, A.M. (1990) Location of transported auxin in etiolated maize shoots using 5-azidoindole-3-acetic acid. *Plant Physiol.* 93: 1154–1161.

Jones, A.M. and Herman, E.M. (1993) KDEL-containing auxin-binding protein is secreted to the plasma membrane and cell wall. *Plant Physiol.* 101: 595–606.

Jones, A.M. and Prasad, P. (1994) The possible function of a nuclear auxin-binding protein. *4th International Congress of Plant Molecular Biology,* Amsterdam. Abstract 900.

Jones, A.M. and Venis, M.A. (1989) Photoaffinity labeling of auxin-binding proteins in maize. *Proc. Natl. Acad. Sci. USA* 86: 6153–6156.

Jones, A.M., Melhado, L.L., Ho, T.H.D. and Leonard, N.J. (1984a) Azido auxins: quantitative binding data in maize. *Plant Physiol.* 74: 295–301.

Jones, A.M., Melhado, L.L., Ho, T.H.D., Pearce, C.J. and Leonard, N.J. (1984b) Azido auxins: photoaffinity labeling of auxin-binding proteins in maize coleoptile with tritiated 5-azidoindole-3-acetic acid. *Plant Physiol.* 75: 1111–1116.

Jones, A.M., Lamerson, P. and Venis, M.A. (1989) Comparisons of site I auxin binding and a 22-kilodalton auxin-binding protein in maize. *Planta* 179: 409–413.

Lazarus, C.M. and Macdonald, H. (1996) Characterization of a strawberry gene for auxin-binding protein, and its expression in insect cells. *Plant Molec. Biol.; in press.*

Lazarus, C.M., Napier, R.M., Yu, L.-X., Lynas, C. and Venis, M.A. (1991) Auxin-binding protein-antibodies and genes. *In:* G.I. Jenkins and W. Schuch (eds): *Molecular Biology of Plant Development.* Company of Biologists, Cambridge, pp 129–148.

Löbler, M. and Klämbt, D. (1985) Auxin-binding protein from coleoptile membranes of corn (*Zea mays* L.). I. Purification by immunological methods and characterization. *J. Biol. Chem.* 260: 9848–9853.

Löbler, M., Simon, K., Hesse, T. and Klämbt, D. (1987) Auxin receptors in target tissues. *In:* J.E. Fox and M. Jacobs (eds), *Molecular Biology of Plant Growth Control.* Alan R. Liss, New York, pp 279–288.

Lomax, T.L. and Hicks, G.R. (1992) Specific auxin-binding proteins in the plasma membrane: receptors or transporters? *Biochem. Soc. Trans.* 20: 64–69.

Macdonald, H., Jones, A.M. and King, P.J. (1991) Photoaffinity labeling of soluble auxin-binding proteins. *J. Biol. Chem.* 266: 7393–7399.

Macdonald, H., Henderson, J., Napier, R.M., Venis, M.A., Hawes, C. and Lazarus, C.M. (1994) Authentic processing and targeting of active maize auxin-binding protein in the baculovirus expression system. *Plant Physiol.* 105: 1049–1057.

Martiny-Baron, G. and Scherer, G.F.E. (1989) Phospholipid-stimulated protein kinase in plants. *J. Biol. Chem.* 264: 18052–18059.

Melhado, L.L., Jones, A.M., Leonard, N.J. and Vanderhoef, L.N. (1981) Azido auxins: synthesis and biological activity of fluorescent photoaffinity labeling agents. *Plant Physiol.* 68: 469–475.

Metzger, H. (1992) Transmembrane signaling: the joy of aggregation. *J. Immunol.* 149: 1477–1487.

Moore, I., Feldwisch, J., Campos, N., Zettl, R., Brzobohaty, B., Baki, L., Schell, J. and Palme, K. (1992) Auxin-binding proteins of *Zea mays* identified by photoaffinity labeling. *Biochem. Soc. Trans.* 20: 70–73.

Murphy, G.J.P. (1980) A reassessment of the binding of naphthaleneacetic acid by membrane preparations of maize. *Planta* 149: 417–426.

Napier, R.M. and Venis, M.A. (1990) Monoclonal antibodies detect an auxin-induced conformational change in the maize auxin-binding protein. *Planta* 182: 313–318.

Napier, R.M. and Venis, M.A. (1992) Epitope mapping reveals conserved regions of an auxin-binding protein. *Biochem. J.* 284: 841–845.

Napier, R.M. and Venis, M.A. (1995) Tansley Review No 79. Auxin action and auxin-binding proteins. *New Phytol.* 129: 167–201.

Napier, R.M., Venis, M.A., Bolton, M.A., Richardson, L.I. and Butcher, D.W. (1988) Preparation and characterization of monoclonal and polyclonal antibodies to maize membrane auxin-binding protein. *Planta* 176: 519–526.

Napier, R.M., Fowke, L.C., Hawes, C., Lewis, M. and Pelham, H.R.B. (1992) Immunological evidence that plants use both HDEL and KDEL for targeting proteins to the endoplasmic reticulum. *J. Cell Sci.* 102: 261–271.

Oliver, S.C., Venis, M.A., Freedman, R.B. and Napier, R.M. (1995) Regulation of synthesis and turnover of maize auxin-binding protein and observations on its passage to the plasma membrane: comparisons to maize immunoglobulin-binding protein cognate. *Planta* 197: 465–474.

Palme, K., Hesse, T., Campos, N., Garbers, C., Yanofsky, M.F. and Schell, J. (1992) Molecular analysis of an auxin binding protein gene located on chromosome 4 of *Arabidopsis. Plant Cell* 4: 193–201.

Pelham, H.R.B. (1989) Control of protein exit from the endoplasmic reticulum. *Annu. Rev. Cell Biol.* 5: 1–23.

Prasad, P.V. and Jones, A.M. (1991) Putative receptor for the plant growth hormone auxin identified and characterized by anti-idiotypic antibodies. *Proc. Natl. Acad. Sci. USA* 88: 5479–5483.

Radermacher, E. and Klämbt, D. (1993) Auxin-dependent growth and auxin-binding proteins in primary roots and root hairs of corn (*Zea mays* L.). *J. Plant Physiol.* 141: 698–703.

Ray, P.M. (1977) Auxin-binding sites of maize coleoptiles are localised on membrane of the endoplasmic reticulum. *Plant Physiol.* 59: 594–599.

Ray, P.M., Dohrmann, U. and Hertel, R. (1977a) Characterization of naphthaleneacetic acid binding to receptor sites on cellular membranes of maize coleoptile tissue. *Plant Physiol.* 59: 357–364.

Ray, P.M., Dohrmann, U. and Hertel, R. (1977b) Specificity of auxin-binding sites on maize coleoptile membranes as possible receptor sites for auxin action. *Plant Physiol.* 60: 585–591.

Rück, A., Palme, K., Venis, M.A., Napier, R.M. and Felle, H. (1993) Patch-clamp analysis establishes a role for an auxin binding protein in the auxin stimulation of plasma membrane current in *Zea mays* protoplasts. *Plant J.* 4: 41–46.

Scherer, G.F.E. (1992) Stimulation of growth and phospholipase A2 by the peptides mastoparan and melittin and by the auxin 2,4-dichlorophenoxyacetic acid. *Plant Growth Regul.* 11: 153–157.

Scherer, G.F.E. (1995) Activation of phospholipase A2 by auxin and mastoparan in hypocotyl segments from zucchini and sunflower. *J. Plant Physiol.* 145: 483–490.

Scherer, G.F.E. and André, B. (1989) A rapid response to a plant hormone: auxin stimulates phospholipase A2 *in vivo* and *in vitro*. *Biochem. Biophys. Res. Comm.* 163: 111–117.

Scherer, G.F.E. and André, B. (1993) Stimulation of phospholipase A2 by auxin in microsomes from suspension-cultured soybean cells is receptor-mediated and influenced by nucleosides. *Planta* 191: 515–523.

Schumaker, K. and Sze, H. (1987) Inositol (1,4,5)trisphosphate releases Ca²⁺ from vacuolar membrane vesicles of oat roots. *J. Biol. Chem.* 262: 3944–3946.

Schwob, E., Choi, S.-Y., Simmons, C., Migliaccio, F., Ilag, L., Hesse, T., Palme, K. and Söll, D. (1993) Molecular analysis of three maize 22 kDa auxin binding protein genes – transient promoter expression and regulatory regions. *Plant J.* 4: 423–432.

Shimomura, S. (1993) EMBL Data Library accession number X70902.

Shimomura, S., Sotobayashi, T., Futai, M. and Fukui, T. (1986) Purification and properties of an auxin binding protein from maize shoot membranes. *J. Biochem.* 99: 1513–1524.

Shimomura, S., Inohara, N., Fukui, T. and Futai, M. (1988) Different properties of two types of auxin-binding sites in membranes from maize coleoptiles. *Planta* 175: 558–566.

Shimomura, S., Liu, W., Inohara, N., Watanabe, S. and Futai, M. (1993) Structure of the gene for an auxin-binding protein and a gene for 7SL RNA from *Arabidopsis thaliana*. *Plant Cell Physiol.* 34: 633–637.

Terryn, N., Van Montagu, M. and Inzé, D. (1993) GTP-binding proteins in plants. *Plant Molec. Biol.* 22: 143–152.

Thiel, G., Blatt, M.R., Fricker, D., White, I.R. and Millner, P. (1993) Modulation of K⁺ channels in *Vicia* stomatal guard cells by peptide homologs to the auxin binding protein C-terminus. *Proc. Natl. Acad. Sci. USA* 90: 11493–11497.

Tillmann, U., Viola, G., Kayser, B., Seimeister, G., Hesse, T., Palme, K., Löbler, M. and Klämbt, D. (1989) cDNA clones of the auxin binding protein from corn coleoptiles (*Zea mays* L.): isolation and characterization by immunological methods. *EMBO J.* 8: 2463–2467.

Venis, M.A. (1977) Solubilisation and partial purification of auxin-binding sites of corn membranes. *Nature* 66: 268–269.

Venis, M.A. (1995) Auxin binding protein 1 is a red herring? Oh no it isn't! *J. Exp. Bot.* 46: 463–465.

Venis, M.A. and Napier, R.M. (1990) Characterization of auxin receptors. *In*: J. Roberts, C. Kirk and M. Venis (eds): *Hormone Perception and Signal Transduction in Animals and Plants*. Company of Biologists, Cambridge, pp 55–65.

Venis, M.A. and Watson P.J. (1978) Naturally occurring modifiers of auxin-receptor interaction in corn: identification as benzoxazolinones. *Planta* 142: 103–107.

Venis, M.A., Thomas, E.W., Barbier-Brygoo, H., Ephritikhine, G. and Guern, J. (1990) Impermeant auxin analogues have auxin activity. *Planta* 182: 232–235.

Venis, M.A., Napier, R.M., Barbier-Brygoo, H., Maurel, C., Perrot-Rechenmann, C. and Guern, J. (1992) Antibodies to a peptide from the auxin-binding protein have auxin agonist activity. *Proc. Natl. Acad. Sci. USA* 89: 7208–7212.

White, I.R., Wise, A. and Millner, P.A. (1993) Evidence for G-protein linked receptors in higher plants: stimulation of GTP-gamma-S binding to membrane fractions by the mastoparan analogue mas 7. *Planta* 191: 285–288.

Wise, A., Thomas, P.G., White, I.R. and Miller, P.A. (1994) Isolation of a putative receptor from *Zea mays* microsomal membranes that interacts with the G-protein, GPα1. *FEBS Lett.* 356: 233–237.

Yu, L.-X. and Lazarus, C.M. (1991) Structure and sequence of an auxin-binding protein gene from maize (*Zea mays* L.). *Plant Mol. Biol.* 16: 925–930.

Zaina, S., Reggiani, R. and Bertani, A. (1990) Preliminary evidence for involvement of GTP-binding protein(s) in auxin signal transduction in rice (*Oryza sativa* L.) coleoptile. *J. Plant Physiol.* 136: 653–658.

Zbell, B. and Walter-Back, C. (1988) Signal transduction of auxin on isolated plant cell membranes: indications for a rapid polyphosphoinositide response stimulated by indoleacetic acid. *J. Plant Physiol.* 133: 353–360.

Zettl, R., Feldwisch, J., Boland, W., Schell, J. and Palme, K. (1992) 5'-azido-[3,6-³H2]-1-naphthylphthalamic acid, a photoactivatable probe for naphthylphthalamic acid receptor proteins from higher plants: identification of a 23-kDa protein from maize coleoptile plasma membranes. *Proc. Natl. Acad. Sci. USA* 89: 480–484.

Zettl, R., Schell, J. and Palme, K. (1994) Photoaffinity labeling of *Arabidopsis thaliana* plasma membrane vesicles by 5-azido-[7-³H]-indole-3-acetic acid: identification of a glutathione S-transferase. *Proc. Natl. Acad. Sci. USA* 91: 689–693.

Signal Transduction in Plants
P. Aducci (ed.)
© 1997 Birkhäuser Verlag Basel/Switzerland

Transduction of ethylene responses

M.A. Hall and A.R. Smith

Institute of Biological Sciences, University of Wales, Aberystwyth, SY23 3DA, UK

Introduction

It has been known for nearly a century (Neljubov, 1901) that ethylene has a wide range of effects on plant growth and development (see e.g., Abeles, 1973). However, with the notable exception of studies on fruit ripening, very little work was performed on the basis of these effects until the early 1960s.

The practical reason for this was that until that time ethylene was the most difficult growth regulator to measure and bioassays were clumsy and laborious. The other principal reason for the slow progress was a reluctance on the part of most people to accept that ethylene – being a gas and a very simple molecule – could possibly be an important natural growth regulator; equally, the technical problems associated with using a gas in experimental work constituted a major disincentive.

However, the advent of gas chromatography made ethylene overnight the easiest growth regulator to measure and progress on the synthesis and mechanism of action in the last 30 years has been considerable and is now probably as well-founded as that for any growth regulator.

On the other hand, while it was recognised that there must exist for ethylene, as for any growth regulator, receptors capable of perceiving and transducing its effects, the technical problems of identifying the perception sites seemed to be insurmountable. Thus, the displacement assays used for example for auxins seemed inappropriate, at least *in vitro*, because unbound ethylene would readily dissipate before it could be measured. At the same time, the very simplicity of the ethylene molecule meant that only relatively low specific activity radioactive ethylene could be obtained, further complicating measurement of low-abundance proteins such as receptors.

As is so often the case, these difficulties did not render such studies impossible, although as we shall see, this owed rather more to serendipity than to foresight.

Early work

Curiously, the earliest studies on the primary mechanism(s) of ethylene action dealt not with classical receptors but rather with ethylene catabolism. Although several microorganisms were known
to possess enzymes capable of metabolising ethylene (de Bont, 1976) it was thought that, unlike
other growth regulators, there was no need for such mechanisms in higher plants since steady-
state ethylene concentrations would be maintained by a balance between synthesis within and
diffusion out of the plant.

Nevertheless, Beyer (1975) was able to demonstrate that higher plants could indeed metabolise
ethylene – to as then unknown products – albeit at very low rates. Rates, indeed, so low that it was
difficult to see how they could have any significant impact on endogenous ethylene concentrations.

However, in subsequent work it was shown that apparent rates of metabolism correlated well
with sensitivity to ethylene in a number of unrelated systems e.g., cotton leaf abscission zones
(Beyer, 1979b), flower fading in carnations (Beyer, 1977), and in morning glory flowers (Beyer
and Sundin, 1978). It was therefore proposed that in fact metabolism was not functional in the
usual way i.e., to control internal concentrations, but rather that the metabolising system had a role
in perception (Beyer, 1979a). This was a departure from the received wisdom concerning receptors for animal hormones, but did not appear to be impossible in principle, especially when the
rather extraordinary nature of the ligand was taken into account. Thus, binding of a ligand to a
receptor or a substrate to an enzyme are not different in kind and both could theoretically bring
about a conformational change initiating a transduction event(s) irrespective of whether the ligand
is subsequently metabolised.

The theory collapsed as a result of various pieces of work. Thus, it was shown that in some
plants (e.g., *Vicia faba*) ethylene was metabolised at high rates and further studies showed that the
product was ethylene oxide (Jerie and Hall, 1978) which was in turn converted to ethylene glycol
and a number of other components. It turned out that this was also the case in the systems exhibiting low rates of metabolism (Blomstrom and Beyer, 1980). Subsequent work showed that in
peas the metabolising systems (probably cytochrome P.450 monooxygenases) were not specific
for ethylene and could indeed metabolise many hydrocarbons (including alkynes, alkanes and
alkenes other than ethylene) often with a higher affinity than that for ethylene (Sanders et al.,
1989a). The final nail in the coffin of the 'metabolism hypothesis' was, however, the demonstration that it was possible to inhibit ethylene metabolism completely with CS_2 without in the least
affecting the plant's ability to respond to ethylene (Sanders et al., 1989a). Almost simultaneously,
ethylene binding sites apparently having the appropriate properties for receptors were defined in
two distinct systems (see below).

Novel analogues of ethylene

One unique aspect of the studies on ethylene binding sites has been the work of Sisler. Thus, although ethylene possesses a wider range of physiologically active analogues with widely differing activities than any other natural plant growth regulator, which in turn enables rigorous characterisation of binding site properties, the approaches available to workers with other putative receptors have appeared to be barred. Thus, while with auxins and gibberellins, the availability of impermeant analogues has enabled innovative approaches in locating and probing for putative receptors, these approaches are impractical for ethylene, since only minor substitutions on the molecule are known to dramatically diminish activity (presumably because of its simplicity and the paucity of recognition points) and attachment of large impermeant molecules such as Sepharose (Hooley et al., 1990) or keyhole limpet haemocyanin (Venis and Napier, 1990) are ruled out.

Sisler (1977) developed the theory that ethylene and other π-acceptor compounds such as carbon monoxide would bind to a metal-containing binding site bringing about a *trans* effect which in turn could alter the conformation of the receptor. To assess this hypothesis Sisler tested a number of π-acceptor compounds not previously thought of as ethylene analogues such as iodide, cyclohexyl isocyanide, phosphorus trifluoride and n-butyl isocyanide which turned out to show high ethylene-like activity in developmental responses such as leaf senescence.

He went on to show that a number of cycloalkenes were capable of inhibiting ethylene responses (Sisler and Pian, 1972; Beggs and Sisler, 1986). A compound in this class which has proved most useful in this area is 2,5-norbornadiene. He also showed that both *cis*-butene (Sisler and Yang, 1984) and *trans* cyclooctene (Sisler et al., 1990) have similar properties.

More recently, Sisler (1996) has shown that methylcyclopropene is an effective inhibitor of ethylene responses and appears to bind irreversibly to the ethylene receptor. This compound in particular opens up tremendous opportunities for future work.

Development of techniques for measuring ethylene binding

Ethylene binding sites were discovered simultaneously and independently in *Phaseolus vulgaris* (Bengochea et al., 1980 a, b) and Sisler (1979). In both cases the affinities of the sites for ethylene and its physiologically active and inactive analogues were exactly as predicted from previous studies on their effects on plant growth and development (Burg and Burg, 1967). Oddly, however, the sites shared another common feature, namely, low rate constants of association and dissociation. This was unexpected since many ethylene effects show short response times (and therefore

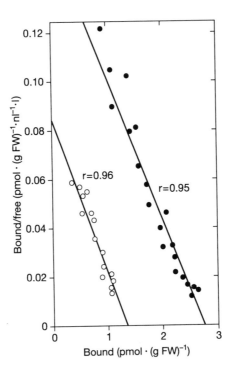

Figure 1. Scatchard plots for ethylene binding to 6-d-old pea epicotyls. Fast associating site (○); slow associating site (●). From Sanders et al. (1991).

are likely to have receptors with high rate constants if the response is proportional to receptor occupancy) but it did prove helpful in purification studies since it was possible to monitor the process by following the distribution of radiolabelled ethylene (see below), which will remain attached to the binding sites for long periods.

In subsequent work, techniques were developed for measuring ethylene binding sites *in vivo* (Sisler and Wood, 1987) the only plant hormone binding site for which this has so far proved possible. These techniques were later modified to take into account effects of ethylene metabolism and endogenous ethylene production (Sanders et al, 1989 a, b).

These studies, which were conducted with a wide range of plant species and tissues sensitive to ethylene, all showed identical features, namely the presence of at least two (or two groups of) sites which, as in the work described above showed, had appropriate affinities for ethylene and its ana logues (Tab. 1, Figs 1 and 2) (Sanders et al., 1991). They differed in one major respect however; thus, one group resembled those from mung bean and *Phaseolus* in having low rate

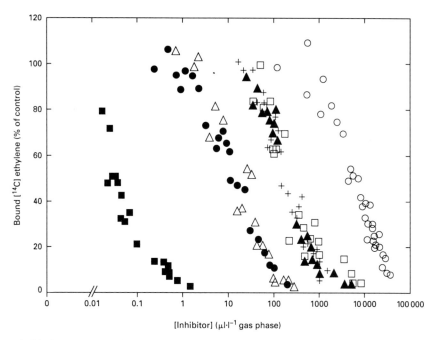

Figure 2. Displacement plots for ethylene and its analogues upon ethylene binding by 6-d-old pea epicotyls. Ethylene (■), propylene (●), acetylene (△), vinyl chloride (▲), methyl acetylene (+), carbon monoxide (□) and 1-butene (○). From Sanders et al. (1991).

Table 1. A comparison of the effects of ethylene analogues and other substances upon ethylene binding and growth of pea epicotyls

Compound	$K_\Theta{}^a$ or K_i for growth inhibition		Partition coefficient	K_D or K_i for binding	
	Gas ($\mu l \times l^{-1}$)	Liquid (M)		Gas ($\mu l \times l^{-1}$)	Liquid (M)
C_2H_4	0.1	$4.6 \cdot 10^{-10}$	0.1	0.0205	$9.1 \cdot 10^{-11}$
C_3H_6	10	$4.9 \cdot 10^{-8}$	1.1	4.6	$2.3 \cdot 10^{-8}$
C_2H_3Cl	140	$5.2 \cdot 10^{-6}$	0.83	56	$2.1 \cdot 10^{-6}$
CO	270	$2.0 \cdot 10^{-7}$	0.017	81	$6.1 \cdot 10^{-8}$
C_2H_2	280	$1.2 \cdot 10^{-5}$	0.95	14.3	$6.1 \cdot 10^{-7}$
C_3H_4	800	$5.4 \cdot 10^{-5}$	1.5	122	$8.2 \cdot 10^{-6}$
$1\text{-}C_4H_8$	27000	$6.9 \cdot 10^{-4}$	0.57	1600	$4.1 \cdot 10^{-5}$
$cis\text{-}2\text{-}C_4H_8$	7100	$1.8 \cdot 10^{-4}$	0.57	5800	$7.5 \cdot 10^{-5}$
2.5-Norbornadiene	42	$1.8 \cdot 10^{-6}$	0.94	28.1	$1.2 \cdot 10^{-6}$
Cyclopentene	3000	$3.88 \cdot 10^{-5}$	0.29	1050	$1.36 \cdot 10^{-5}$
CO_2	150000	$5.7 \cdot 10^{-4}$	0.83	NE	-
$AgNO_3$	-	$1.42 \cdot 10^{-4}$	-	-	NE

^{a} K_Θ represents the concentration of ligand giving a half-maximal response.
The figures given are those for binding sites with high rate constants of association, those for sites with low rate constants are not significantly different. From Sanders et al. (1991)

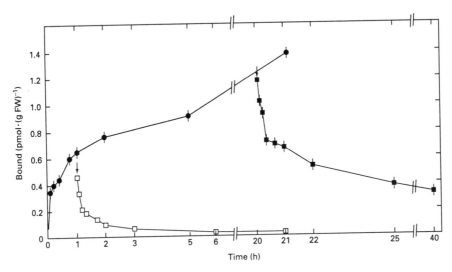

Figure 3. Association and dissociation plots for ethylene binding to 6-d-old pea epicotyls. Association (●), dissociation (□, ■). Saturating concentrations of $^{12}C_2H_4$ were added at the arrows, to produce the dissociation plots. From Sanders et al. (1991).

constants of association and dissociation but the other showed high rate constants (Sanders et al., 1990, 1991a, b). In effect, the time to half-maximum association for one group was of the order of tens of hours while that of the other was measured in minutes (Fig. 3). In contrast to the situation in *Phaseolus* cotyledons and mung beans the concentration of both groups of sites was very low (1–2 pmol $\times g^{-1}$ f.wt.) in all species and tissues examined.

Distribution of binding sites

The only work on the subcellular distribution of binding sites has been performed in *Phaseolus* cotyledons. Using high-resolution EM autoradiography on *Phaseolus* tissue where bound ^{14}C ethylene had been fixed with osmium tetroxide which cross links the double bond in ethylene, sites were located on protein body membranes and the membranes of the endoplasmic reticulum (Evans et al., 1982a) Parallel work using marker enzymes led to a similar conclusion with the additional possibility of a small population being present on the plasmalemma (Evans et al., 1982b).

There is much less information on tissue distribution. In pea epicotyls the most ethylene- sensitive tissues close to the apex show a higher concentration of both fast and slow associating sites

Table 2. Distribution of ethylene binding sites in pea epicotyls

Tissue location	Concentration of ethylene binding sites (pmol×g⁻¹ f.wt.)		Number of ethylene binding sites per cell (×10⁻⁵)	
	Fast associating	Slow associating	Fast associating	Slow associating
Tip	3.12	3.86	1.96	2.42
Remainder of second internode	0.58	1.26	1.90	4.13
First internode	0.44	0.96	1.77	3.87

Adapted from Sanders et al. (1991); f.wt. = fresh weight

than the less sensitive tissue below but, curiously, there is an almost identical amount of both sites per cell down the whole length of the epicotyl (Tab. 2) (Sanders et al., 1991). This indicates either a very strict control of turnover or that the complement of sites is set early in ontogeny and turnover is negligible. Similar results were obtained with rice coleoptiles where the concentration of sites falls in direct proportion to the degree of cell extension (cell division does not occur at this stage in this tissue) (Sanders et al., 1990). Other work using either direct measurements or antibodies raised to the ethylene binding site from *Phaseolus* indicated that there were higher concentrations of binding protein in abscission zones, an ethylene sensitive tissue, than in adjacent petiolar tissues (Connern et al., 1989). Concentrations of binding sites (both groups) were very similar in elongating tissues whether these respond to ethylene through inhibition or through promotion.

One further interesting feature is that in peas it has proved possible to detect ethylene binding sites of both groups *in vitro* although the amounts extracted are close to the limits of detection. In endomembrane preparations both groups of sites are detectable, but in cytosolic fractions only sites with high rate constants of association and dissociation appear to be present (Moshkov et al., 1993). This may not however reflect a true distribution but rather a weaker association of such sites with a membrane than that obtained with the strongly membrane bound sites having low rate constants (see below).

Separation and purification of binding sites

Early work indicated that the sites both in mung bean and in *Phaseolus* were membrane-bound (Bengochea, 1980a) and it soon emerged that they were highly hydrophobic and could only be solubilised by the use of detergents. This has proved to be a major obstacle in subsequent purification work. The earliest studies on the *Phaseolus* protein yielded a partial purification and gave a value for the molecular weight of the order of 50–60 kD using gel permeation chromatography

(Thomas et al., 1984). Such estimates are however fraught with difficulty in the case of hydrophobic proteins since these can bind up to many times their own weight in detergent with a consequent apparent increase in molecular dimensions. The advent of FPLC gave added impetus to the work and substantial further purification, but electrophoresis on native gels was unsatisfactory because of the presence of detergent and denaturing gels showed only low molecular weight fragments (10–12 kDa). Moreover, under such conditions all bound radioactivity was lost and such fragments themselves did not bind ethylene. The use of semi-denaturing conditions (Harpham et al., 1995) however allowed the separation of two bands located by bound $^{14}C_2H_4$ at 28 and 26 kDa, *both* of which bind ethylene; the 28 kDa band appears to be glycosylated. The purity of the proteins was confirmed by 2-D electrophoresis.

Antibodies to both the bands have been used to screen Lambda ZAP expression libraries from *Phaseolus* abscission zones and *Arabidopsis* leaves but no positive results were obtained, probably due to a combination of low antibody avidity and the difficulty of detecting a low abundance protein. The antibodies also recognise a protein with a molecular mass of around 56 kDa on semi-denaturing gels.

N-terminal and internal sequences have been obtained for the 28kD band amounting to about 25% of the total. A 16 amino acid N-terminal sequence shows no homology to other sequences in the Swiss Prot databases. However, one of the internal sequences shows 75% identity in an eight amino acid overlap to a protein kinase substrate (Harpham et al., 1996), which may have relevance to the transduction studies outlined below.

Antibodies are now being raised to synthetic peptides based on two of the internal sequences and will be used to screen for homologous peptides in a range of systems.

Molecular genetics

Work in this area is of a relatively recent date but has resulted in some dramatic progress. Bleecker et al. (1988) were the first to publish information on the *Arabidopsis* mutant *etr* which shows almost complete insensitivity to ethylene and where, moreover, ethylene binding is significantly reduced. Subsequently, other mutants were produced showing properties similar to *etr* namely *ein* (Guzman and Ecker, 1990) and *eti* (Harpham et al., 1991), – but also others showing a constitutive ethylene phenotype – *eto*, *ctr* (Kieber et al., 1993) and *his* 1 (Guzman and Ecker, 1990). (Figs 4 and 5).

Studies on ethylene binding in the *eti* series showed a complex situation. *eti* 5 resembles *etr* in being completely insensitive to ethylene and has reduced concentrations of fast associating site compared to the wild type but similar concentrations of slow associating sites (Sanders et al., 1991). (Tab. 3). However, the rest of the series show increased or decreased concentrations of fast associating site coupled with similar or reduced concentrations of slow associating site. Some, at least, of these mutants are probably transduction rather than perception mutants but a word of caution is appropriate since measurements of ethylene binding do not necessarily reflect total amounts and functionality of the binding protein, either because the lesion is in a factor affecting the affinity of the binding domain for ethylene (see below), or because the lesion may be in part of the binding protein which does not affect the binding domain. Equally, because all the measurements have, of necessity, been performed *in vivo*, it is vital to take into account the effects of ethylene biosynthesis (Sanders et al., 1991) especially since mutants may differ significantly in this respect.

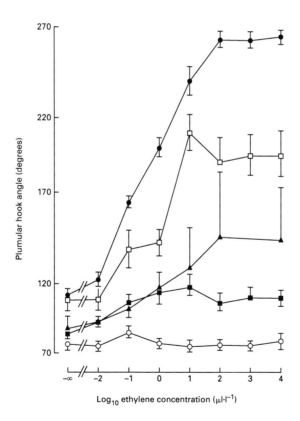

Figure 4. The effect of ethylene upon plumular hook angle in wild type and mutant *Arabidopsis*. Wild type (●), *eti* 3 (□), *eti* 5 (○), *eti* 10 (■), *eti* 13 (▲). From Harpham et al. (1991).

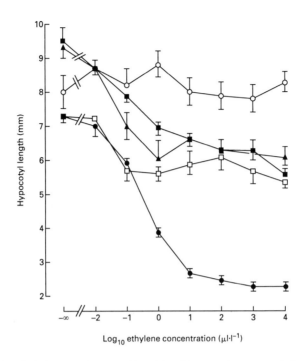

Figure 5. The effect of ethylene upon hypocotyl length in wild type and mutant *Arabidopsis*. Wild type (●), *eti* 3 (□), *eti* 5 (○), *eti* 10 (■), *eti* 13 (▲).

One further striking feature of the mutants showing a complete lack of sensitivity to ethylene is that feedback inhibition of ethylene biosynthesis is abolished (Sanders et al., 1991). This is in accord with much previous work which indicates that such feedback inhibition is receptor mediated (Wang and Woodson, 1989).

Table 3. Ethylene binding in wild-type and mutant plants of *Arabidopsis*

	General order of sensitivity to ethylene	Concentration of fast associating sites (pmol×g⁻¹ f.wt.)	Concentration of slow associating sites (pmol×g⁻¹ f.wt.)
Wild type	1	0.66	0.15
eti 3	2	0.76	-
eti 5	6	0.28	1.12
eti 8	3=	0.39	-
eti 10	5	0.73	0.12
eti 13	3=	0.89	-

Adapted from Sanders et al. (1991); f.wt. = fresh weight

One curious feature which distinguishes all the ethylene sensitivity mutants so far described from, for example, auxin sensitivity mutants, is that while in the case of the latter, sensitivity to auxin is decreased by one or two orders of magnitude (i.e., the maximal response may be restored if a sufficiently high concentration of auxin is applied (Barbier-Brygoo et al., 1990)), this is not the case for the ethylene mutants where increasing ethylene concentrations even by four or five orders of magnitude either has no effect whatever (e.g., *eti* 5 (Harpham et al. 1991)) or does not restore the maximal (i.e., wild type) response. This appears to signify that the mutations are different in kind to those seen with auxin-sensitivity mutants. Certainly in the case of *etr* the mutation appears to be in the receptor itself (see below) whereas in the case of auxin receptors the lesion appears to affect receptor abundance at the plasmalemma.

The mutant *ctr* 1 displays a constitutive triple response which is not reversed by inhibitors of ethylene biosynthesis or action suggesting that it is defective in ethylene signal transduction (Kieber et al., 1993). The *CTR* 1 gene has been cloned and its sequence has the characteristics of a threonine-serine protein kinase (Kieber et al., 1993) particularly of the Raf family which is involved in a variety of developmental events in eukaryotes. The same work indicated that the *CTR*1 gene product acts downstream of *ETR* 1 and *EIN* 4 and is a negative regulator of several of the *EIN* genes. Transformation of *Arabidopsis* with the *ETR* gene confers ethylene insensitivity (Chang et al., 1993), which is not perhaps surprising. However, cloning of the gene by chromosome walking (Chang et al., 1993) showed that while N-terminal sequences had no similarity to sequences in available databases, the remaining carboxyl-terminal sequence is very similar to the conserved domains of both the sensor and response regulator domains in the prokaryotic two-component system of signal transduction. The size of protein encoded by the *ETR* gene is of the order of 80 kDa, but other work indicates that it may exist *in vivo* as a disulphide-linked dimer (Schaller et al., 1995).

In such two-component systems, signal perception by the N-terminal domain (which is usually located in the periplasmic space flanked by two or more transmembrane domains) results in auto-phosphorylation of a carboxyl-terminal histidine kinase domain; the phosphate group is then transferred to an aspartate residue in the N-terminal domain of the response regulator which in turn regulates its activity – for example as a transcriptional activator (Stock et al., 1985).

An exciting development of this work has demonstrated that transformation of yeast with the wild type *ETR* gene conferred the ability to bind ethylene in a saturable and reversible manner, with an appropriate K_D (Schaller and Bleecker, 1995). This comes very close to proving that the *ETR* gene product is indeed an ethylene receptor, although, allowing for the complexity of the system, further molecular genetical and biochemical work is necessary. One curious feature of this work is that when ethylene binding in *etr* was found to be lower than that in wild type, an incubation period of 6 h was used. This implies that the deficiency was in fast-associating sites at least. However when *ETR* was expressed in yeast the $t_{1/2}$ for association was shown to be 12.5 h which corresponds to the figure for slow-associating sites.

Very recently a gene named *ERS* has been isolated from *Arabidopsis* by cross-hybridisation with the *ETR 1* gene (Hua et al., 1995). The deduced ERS protein shows significant homology with the N-terminal and putative histidine protein kinase domains of ETR 1 but lacks the receiver domain of the latter. Transformation of *Arabidopsis* with a missense mutated ERS gene conferred ethylene insensitivity. This is the first evidence that a family of such genes exists.

In this connection Yen et al. (1995) have shown that the Never-ripe (Nr) locus in tomato (Nr fruits are insensitive to ethylene) may be homologous to the *ETR* gene. Wilkinson et al. (1995) have also shown that expression of the normal NR gene is itself modulated by ethylene.

Transduction events post-perception

The first biochemical indications that protein kinase activity was involved in the transduction of ethylene responses was the demonstration that application of the gas to pea tissue or endomembrane preparations from such tissue led to marked increases in protein phosphorylation in both soluble and membrane-bound fractions (Novikova et al., 1993).

The extent of this phosphorylation was dose-dependent and the effect was abolished by the ethylene antagonist norbornadiene. Of special interest was the finding that enhanced phosphorylation occurred in a zone on semi-denaturing gels corresponding to that occupied by ethylene binding activity. Using antibodies raised to the 26 and 28 kD proteins from *Phaseolus* it was shown that immunoprecipitated proteins had been phosphorylated in response to ethylene treatment.

In parallel work with membrane-enriched fractions, it was demonstrated that under conditions where protein kinase activity was inhibited, ethylene binding activity increased, whereas the converse was true where phosphorylation events were promoted (Novikova et al., 1993) Okadaic acid, an inhibitor of serine threonine protein phosphatases 2 A and B, also inhibits ethylene production in *Arabidopsis*. Equally, while lanthanum chloride, a calcium channel blocker, promotes ethylene biosynthesis in wild type plants it is ineffective with *eti* 5, where, as suggested above, receptor-mediated control of ethylene biosynthesis is absent (Berry et al., 1996).

Simultaneously Raz and Flur (1993) showed that in tobacco, ethylene application induced rapid and transient protein phosphorylation which was blocked by protein kinase inhibitors and enhanced by okadaic acid. Strikingly, the induction of pathogenesis related (*PR*) genes by ethylene in the same system paralleled the effects of the inhibitors i.e., protein kinase inhibitors abolished the induction of *PR* genes while okadaic acid enhanced their induction – even in the absence of ethylene.

Interestingly, but perhaps coincidentally, two of the proteins showing transitory phosphorylation in tobacco leaves have similar molecular masses to the ethylene binding sites from *Phaseolus.*

In a somewhat different context, similar results have been obtained using the ethylene-induced triple response in peas (Berry et al., 1996). Both the protein kinase inhibitor staurosporine and okadaic acid reverse ethylene-induced inhibition of elongation, and while this again indicates a role for protein phosphorylation in the transduction of ethylene responses, it must be remembered that other growth regulators, particularly auxins in this case, are also involved and their effects may also involve protein kinase cascades. Indeed, this may account for a lack of effect of the inhibitors on the other two components of the triple response, namely hook closure and isodiametric cell expansion. Porat et al. (1994) have shown that in *Phalaenopsis,* protein phosphorylation increased significantly following pollination, a process which increases sensitivity to ethylene. Treatment with Ca^{2+} (a co-factor of many protein kinases) increased sensitivity to ethylene while treatment with the chelator EGTA has the reverse effect. However, the ethylene modulated phosphorylation in membrane fractions from peas appears to be dependent on the presence of Mg^{2+} and not Ca^{2+}. Further studies in our laboratory on protein phosphorylation using the *eti* series of mutants indicates that in *eti* at least protein phosphorylation is enhanced both quantitatively and qualitatively. This is in accord with the studies on *etr* and other mutants indicating that protein kinase cascades are part of the signaling pathway.

Very recently it has been shown in our laboratory that both in peas and in *Arabidopsis* ethylene treatment results in the phosphorylation of a protein which cross-reacts with antibodies raised to a nucleotide diphosphate kinase from peas. The significance of this finding in unclear but may be related to the studies on GTP-binding proteins outlined below.

While much of the evidence using inhibitors is circumstantial, that work, together with the data on protein phosphorylation *in vivo* and *in vitro*, together with the observed effects on ethylene binding itself, do point in the same direction as the findings with molecular genetics; namely, that protein phosphorylation cascades are involved in the transduction of ethylene effects and indeed may be involved centrally in regulation at the receptor level as in the case of β-adrenergic receptors in animals (Hausdorff et al., 1990). In studies on senescence in *Phalaenopsis* flowers, Porat et al. (1994) noted that cholera toxin and GTPγS increased sensitivity to ethylene, indicating an involvement of G-proteins in this process. Equally, 10 h after pollination, at the peak of ethylene sensitivity, increased binding of GTP to microsomal membranes was observed.

In our laboratory it has been shown that partially purified fractions from *Phaseolus* cotyledons containing ethylene binding proteins bind [α^{32}P]GTP used as an affinity probe (Löw et al., 1992). Such binding is significantly reduced in the presence of non-radioactive GTP and GDPβS. On semi-denaturing gels a band at about 23 kDa showed the greatest binding activity.

Similarly, in membrane preparations from pea epicotyls specific [^{35}S] GTPγS binding was detected with a K_d of about 10^{-10} M. Studies using [α^{32}P] GTP showed essentially the same response. Moreover, solubilised membrane preparations from tissue treated with ethylene for 1 h showed increased specific GTP binding to one band at 23 kDa predominating. Mastoparan had no effect on GTPγS binding in this system and in any case the band at 23 kDa does not correspond with the molecular mass of any other subunit of a heterotrimeric G-protein.

On the other hand, the labelled 23 kDa band coaligned with a band on Western blots which were prepared using antibodies to pan-Ras which corresponds to one of four consensus sequences involved with GTP binding in the Ras group of proteins.

These findings provide strong circumstantial evidence that G-proteins are involved with the transduction of ethylene responses. Similar work using *Arabidopsis* gave identical results except that in this case at least two small GTP-binding proteins are activated by ethylene, moreover the synthetic cytokinin benzylaminopurine antagonised this effect *in vitro*. Moreover, in *eti* 5 both proteins showed higher activity than wild type irrespective of hormone treatment.

Conclusions

While it is clear from the foregoing that enormous strides have been made in the elucidation of mechanisms of ethylene perception and transduction since the discovery of binding sites 15 years ago, there nonetheless remains much that is obscure. In part at least, this is due to the complexity of the systems which are emerging and which might have been expected of a plant growth regulator exhibiting pleiotropic effects. Signaling pathways in animals show great complexity and, given the paucity of known plant hormones, it is not to be expected that the situation with these will be less complex.

Clearly, the *ETR* gene product appears to be a strong candidate for an ethylene receptor but equally on the basis of size alone, this does not code for the same proteins as those found in *Phaseolus*, although this difference may be more apparent than real and may relate either to differences in electrophoretic conditions or to fragmentation of the *Phaseolus* protein during purification.

It can be argued that this indicates that the protein in *Phaseolus* is not in fact a receptor, but specificity and ubiquity (since kinetically they correspond to the sites with low rate constants of association and dissociation found in the many different species examined so far) makes this highly unlikely, especially since we know that they are not ethylene metabolising enzymes. However, the demonstration that when the *ETR* gene is expressed in yeast, ethylene binding shows the characteristics of slow associating sites even though the gene was selected on the basis of a deficiency in fast-associating sites suggests strongly that these proteins are one and the same thing or at least isoforms and that the differences in rate constants reflect differences in cellular location or post-transcriptional modification.

In respect of transduction responses the story is somewhat clearer, at least in principle. Both approaches indicate an important involvement of protein kinases, whether this results in phosphorylation of the receptor itself or of other proteins or both. The recent work in peas and *Arabidopsis* on activation of small GTP binding proteins by ethylene strongly suggests a place for these close to the receptor or even as the output domain of the response regulator. In this connection it is perhaps significant that *ras*-type proteins control Raf kinases to which family the *CTR* gene belongs. This is an area where rapid future progress can be foreseen.

A major drawback in all these studies has been the lack of an adequate functional assay for ethylene receptors unlike the elegant demonstrations with auxin receptors (see Venis and Napier, this volume). It is an urgent requirement that such be developed since none of the work outlined above will be a satisfactory demonstration of receptor capability until the genetical and biochemical results have been incontrovertibly linked to a physiological response. Equally, only by an integrated approach involving physiological, biochemical and molecular work are the complex problems evidenced here likely to be resolved.

References

Abeles, F.B. (1973) *Ethylene in Plant Biology*. Academic Press. New York.

Barbier-Brygoo, H., Maurel, C., Shen, W.H., Ephritikine, G., Delbarre, A and Guern, J. (1990) Use of mutants and transformed plants to study the action of auxins *In*: Roberts, J., Kirk, C. and Venis, M.A. (eds): *Hormone Perception and Signal Transduction in Animals and Plants*. Company of Biologists, Cambridge. pp 67–78.

Beggs, M.J. and Sisler, E.C. (1986) Binding of ethylene analogs and cyclic olefins to a Triton X-100 extract from plants: Comparison with *in vivo* activities. *Plant Growth Regul.* 4: 13–21.

Bengochea, T., Dodds, J.H., Evans, D.E., Jerie, P.H., Niepel, B., Shaari, A.R. and Hall, M.A. (1980a) Studies on ethylene binding by cell-free preparations from cotyledons of *Phaseolus vulgaris* L. I Separation and characterisation. *Plants* 148: 397–406.

Bengochea, T., Acaster, M.A., Dodds, J.H., Evans, D.E., Jerie, P.H. and Hall, M.A. (1980b) Studies on ethylene binding by cell-free preparations from cotyledons of *Phaseolus vulgaris* L. II: Effects of structural analogues of ethylene and of inhibitors. *Plants* 148: 407–411.

Berry, A.W., Cowan, D.S.C., Harpham, N.V.J., Hemsley, R.J., Novikova, G.V., Smith, A.R. and Hall, M.A. (1996) Studies on the possible role of protein phosphorylation in the transduction of the ethylene signal. *Plant Growth Regul.* 18: 135–141.

Beyer, E.M. (1975) $^{14}C_2H_4$: Its incorporation and metabolism by pea seedlings under aseptic conditions. *Plant Physiol.* 56: 273–278.

Beyer, E.M. (1977) $^{14}C_2H_4$: Its incorporation and oxidation to $^{14}CO_2$ by cut carnations. *Plant Physiol.* 60: 203–206.

Beyer, E.M. (1979a) Effect of silver ion, carbon dioxide and oxygen on ethylene action and metabolism. *Plant Physiol.* 63: 169–173.

Beyer, E.M. (1979b) [^{14}C] ethylene metabolism during leaf abscission in cotton. *Plant Physiol.* 64: 971–974.

Beyer, E.M. and Sundin, O. (1978) $^{14}C_2H_4$ metabolism in morning glory flowers. *Plant Physiol.* 61: 896–899.

Bleecker, A.B., Estelle, M.A., Somerville, A.C. and Kende, H. (1988) Insensitivity to ethylene conferred by a dominant mutation in *Arabidopsis thaliana. Science* 241: 1086–1089.

Blomstrom, D.C. and Beyer, E.M. (1980) Plants metabolise ethylene to ethylene glycol. *Nature* 233: 66–67.

Burg, S.P. and Burg, E.A. (1967) Molecular requirements for the biological action of ethylene. *Plant Physiol.* 44: 144–152.

Chang, C., Kwok, S.F., Bleecker, A.B. and Meyerowitz, E.M. (1993) *Arabidopsis* ethylene response gene ETR 1: Similarity of product to two-component regulators. *Science* 262: 539–544.

Connern, C.P., Smith, A.R., Turner, R. and Hall, M.A. (1989) Putative ethylene binding proteins from abscission zones of *Phaseolus vulgaris. In:* D.J. Osborne and Jackson, M.B. (eds): *Cell Separation in Plants* (NATO ASI Series, Vol. H35) Springer-Verlag, Berlin, Heidelberg, pp 351–356.

de Bont, J.A.M. (1976) Oxidation of ethylene by soil bacteria. *Anton. Leeuwenhoek Int. J. Gen. M.* 42: 59–71.

Evans, D.E., Dodds, J.H., Lloyd, P.C. ap Gwynn, I. and Hall, M.A. (1982a) A study of the subcellular localisation of an ethylene binding site in developing cotyledons of *Phaseolus vulgaris* L. by high resolution autoradiography. *Planta* 154: 48–52.

Evans, D.E., Bengochea, T., Cairns, A.J., Dodds, J.H. and Hall, M.A. (1982b) Studies on ethylene binding by cell-free preparations from cotyledons of *Phaseolus vulgaris* L.: Subcellular localisation *Plant Cell Environ.* 5: 101–107.

Guzman, P. and Ecker, J.R. (1990) Exploiting the triple response of *Arabidopsis* to identify ethylene-related mutants. *Plant Cell* 2: 513–523.

Hausdorff, W.P., Lohse, M.J., Bouvier, M., Liggett, S.B., Caron, M.G. and Lefkowitz, R.J. (1990) Two kinases mediate agonist-dependent phosphorylation and desensitization of the β_2-adrenergic receptor *In*: Roberts, J. Kirk, C. and Venis, M.A. (eds): *Hormone Perception and Signal Transduction in Animals and Plants.* The Company of Biologists, Cambridge, pp 225–240.

Harpham, N.V.J., Berry, A.W., Knee, E.M., Roveda-Hoyos, G., Raskin, I., Sanders, I.O., Smith, A.R., Wood, C.K. and Hall, M.A. (1991) The effect of ethylene on the growth and development of wild-type and mutant *Arabidopsis thaliana* (L.) Heynh. *Annu. Bot.* 68: 55–61.

Harpham, N.V.J., Berry, A.W., Holland, M.G., Moshkov, I.E., Smith, A.R. and Hall, M.A. (1996) Ethylene binding sites in higher plants. *Plant Growth Regul.* 18: 71–77.

Hooley, R. Beale, M.H. and Smith, S.J. (1990) Gibberellin perception in the *Avena fatua* aleurone. *In*: Roberts, J., Kirk, C. and Venis, M.A. (eds): *Hormone Perception and Signal Transduction in Animals and Plants.* Company of Biologists, Cambridge, pp 79–86.

Hua, J., Chang, C., Sun, Q. and Meyerowitz, E.M. (1995) Ethylene insensitivity conferred by *Arabidopsis* ERS gene. *Science* 269: 1712–1713.

Jerie P. H. and Hall, M.A. (1978) The identification of ethylene oxide as a major metabolite of ethylene in *Vicia faba* L. *Proc. R. Soc. Lond. B.* 200: 87–94.

Kieber, J.J., Rothenberg, M., Roman, G., Feldman, K.A. and Ecker, J.R. (1993) CTR1, a negative regulator of the ethylene response pathway in *Arabidopsis,* encodes a member of the *raf* family of protein kinases. *Cell* 72: 427–441.

Löw, A., Faulhanmer, H.G. and Sprinzl, M. (1992) Affinity labeling of GTP-binding proteins in cellular extracts. *FEBS Lett.* 303: 64–68.

Moshkov, I. Ye., Novikova, G.V., Smith, A.R. and Hall, M.A. (1993) *In vitro* study of ethylene binding sites in pea seedlings. *In*: Pech, J.C., Latche, A. and Balague (eds): *Cellular and Molecular Aspects of the Plant Hormone Ethylene.* Kluwer Academic Publishers, London, pp 195–196.

Neljubov, D. (1901) Über die horizontale Nutation der Stengel von *Pisum sativum* und einiger anderer. Pflanzen *Beih. Bot. Zentralbl.* 10: 128–139.

Novikova, G.V., Moshkov, I.E., Smith, A.R. and Hall, M.A. (1993) Ethylene and phosphorylation of pea epicotyl proteins. *In*: Pech, J.C., Latche, A. and Balague, (eds): *Cellular and Molecular Aspects of the Plant Hormone Ethylene.* Kluwer Academic Publishers, London, pp 371–372.

Porat, R., Borochov, A. and Halevy, A.H. (1994) Pollination-induced senescence in phalaenopsis petals. Relationship of ethylene sensitivity to the activity of GTP-binding proteins and protein phosphorylation. *Physiol. Plant.* 90: 679–684.

Raz, V. and Fluhr, R. (1993) Ethylene signal is transduced *via* protein phosphorylation events. *Plant Cell* 5: 523–530.

Sanders, I.O., Smith, A.R. and Hall, M.A. (1989a) Ethylene metabolism in *Pisum sativum* L. *Planta* 179: 104–114.

Sanders, I.O., Smith, A.R. and Hall, M.A. (1989b) The measurement of ethylene binding and metabolism in plant tissue. *Planta* 179: 97–103.

Sanders, I.O., Ishizawa, K., Smith, A.R. and Hall, M.A. (1990) Ethylene binding and action in rice seedlings. *Plant Cell Physiol.* 31 (8): 1091–1099.

Sanders, I.O., Smith, A.R. and Hall, M.A. (1991a) Ethylene binding in *Pisum sativum*. *Planta* 183: 209–217.

Sanders, I.O., Harpham, N.V.J., Raskin, I., Smith, A.R. and Hall, M.A. (1991b) Ethylene binding in wild type and mutant *Arabidopsis thaliana* (L.) *Heynh*. *Planta* 68: 97–103.

Schaller, G.E. and Bleecker, A.B. (1995) Ethylene-binding sites generated in yeast expressing the *Arabidopsis* ETR1 gene. *Science* 270: 1809–1811.

Schaller, G.E., Ladd, A.N., Lanahan, M.B., Spanbauer, J.M. and Bleecker, A.B. (1995) The ethylene response mediator ETR 1 from *Arabidopsis* forms a disulfide-linked dimer. *J. Biol. Chem.* 270: 12526–12530.

Sisler, E.C. (1977) Ethylene activity of some π-acceptor compounds. *Tob. Sci.* 21: 43–45.

Sisler, E.C. (1979) Measurement of ethylene binding in plant tissue. *Plant Physiol.* 64: 538–542.

Sisler, E.C. (1996) Effect of 1-methylcyclopropene and methylenecyclopropane on ethylene binding and ethylene action on cut carnations. *Plant Growth Regul.* 18: 79–86.

Sisler, E.C. and Pian, A. (1972) Effect of ethylene and cyclic olefins on tobacco leaves. *Tob. Sci.* 17: 68–72.

Sisler, E.C. and Wood, C. (1987) Ethylene binding and evidence that binding *in vivo* and *in vitro* is to the physiological receptor. In: D. Klämbt (ed.): *Plant Hormone Receptors* (NATO ASI Series, Vol. H10) Springer–Verlag, Berlin, Heidelberg, New York, pp 239–248.

Sisler, E.C. and Yang, S.F. (1984) Anti-ethylene effects of *cis*-2-butene and cyclic olefins. *Phytochemistry* 23: 2765–2768.

Sisler, E.C., Blankenship, S.M. and Guest, M. (1990) Competition of cyclooctenes and cyclooctadienes for ethylene binding and activity in plants. *Plant Growth Regul.* 9: 157–164.

Stock, A., Koshland, D.E.J., Jr. and Stock, J. (1985) Homologies between the *Salmonella typhimurium* Che Y protein involved in the regulation of chemotaxis, membrane protein synthesis and sporulation. *Proc. Natl. Acad. Sci. USA* 82: 7989–7993.

Thomas, C.J.R., Smith, A.R. and Hall, M.A. (1984) The effect of solubilisation on the character of an ethylene binding site from *Phaseolus vulgaris* cotyledons. *Planta* 160: 474–479.

Venis, M.A. and Napier, R.M. (1990) Characterisation of auxin receptors. In: Roberts, J., Kirk, C. and Venis M.A. (eds): *Hormone Perception and Signal Transduction in Plants and Animals*. Company of Biologists, Cambridge, pp 55–66.

Wang, H. and Woodson, W.R. (1989) Reversible inhibition of ethylene action and interruption of petal senescence in carnation flowers by norbornadiene. *Plant Physiol.* 89: 434–438.

Wilkinson, J.Q., Lanahan, M.B., Hsiao-Ching Yen, Giovannoni, J.J., Klee, H.J. (1995) An ethylene-inducible component of signal transduction encoded by *Never-ripe*. *Science* 270: 1807–1809.

Yen, H.C., Sanghyeob, L. Tanksley, S.D., Lanahan, M.B., Klee, H.J. and Giovannoni, J.J. (1995) The tomato *Never-ripe* locus regulates ethylene inducible gene expression and is linked to a homolog of the *Arabidopsis* ETR 1 gene. *Plant Physiol.* 107: 1343–1353.

Signal Transduction in Plants
P. Aducci (ed.)
© 1997 Birkhäuser Verlag Basel/Switzerland

Phytotoxins as molecular signals

P. Aducci[1], A. Ballio[2] and M. Marra[1]

[1]Department of Biology, University of Rome "Tor Vergata", I-00133 Rome, Italy
[2] Department of Biochemical Sciences "Alessandro Rossi-Fanelli", University of Rome "La Sapienza", I-00185 Rome, Italy

Introduction

It is customary to include under the name of phytotoxins those microbial metabolites that, with the exclusion of enzymes, damage plants at low concentrations. Many plant pathogenic bacteria and fungi produce phytotoxins in culture, but the potential role of these metabolites in pathogenesis has been seldom demonstrated. Essential conditions for attributing the function of disease determinants are: a) the demonstration that the toxin occurs in infected plants, b) the ability of the toxin to reproduce at least a portion of the syndrome when placed in a healthy plant.

Several dozens of phytotoxins have been isolated and their structure elucidated; they show a very great chemical diversity. The molecular mechanisms responsible for their toxicity have been investigated in a limited number of cases, both in the group of the so-called host-selective toxins, namely those which reproduce the symptoms of the disease only in genotypes of the host that are susceptible to the pathogen, and in non-selective toxins, which do not reproduce the pattern of resistance and susceptibility of the host to the pathogen. The investigations on the mode of action of phytotoxins, necessary to prove the nature of the molecular lesion, have rarely penetrated into the biochemical machinery of the cell. Consequently, the information about perception is very limited, and even more so the knowledge of the steps involved in the transduction of signals to the target and in the final expression of symptoms. Scheffer and Pringle (1964) postulated the existance of specific receptors for a phytotoxin in susceptible host tissues. But since then the characterization of binding sites in phytotoxins has made very little progress. Among other reasons, this might be due to the frequently encountered difficulty in the preparation of a radioactive derivative which combines high specific activity with retention of the biological properties proper to the original metabolite.

The following discussion deals with some recent findings more or less directly related to the problem of phytotoxin recognition and intracellular signaling in the host. The authors have preferred to restrict the discussion to a limited number of representative examples, all chosen among non-host-selective toxins of fungal origin, rather than broadly surveying the subject. The non-selectivity of several phytotoxins, evidenced by the lesion of some physiological processes in a large number of hosts, depends on the interference with a precise biochemical target of wide, if not ubiquitous, occurrence in plants. Whenever the target corresponds to a crucial step in signal transduction the toxin can represent a useful tool for the search for a plausible signaling cascade. So far, evidence that secondary messengers are involved in the events following perception of a phytotoxin is, at best, only circumstantial.

Beticolins and cebetins

The fungus *Cercospora beticola* is the causal agent of the sugar beet leaf spot disease. It produces two types of host-nonselective phytotoxins suspected of being involved in the development of disease symptoms.

The first toxic metabolite is cercosporin (Weiss et al., 1987), a red perilenquinone derivative found also in numerous other species of *Cercospora* (Assante et al., 1977). The damage to the plant cell is caused by the formation on illumination of highly toxic singlet oxygen responsible for lipid peroxidation and consequently for the loss of membrane integrity (Macrì and Vianello, 1979a). This photoactivatable toxin has become a useful tool in fungal photobiology (Daub and Ehrenshaft, 1993).

The second type of toxic metabolites comprehends several yellow photoactivatable compounds called beticolins (Milat et al., 1992b; Milat et al., 1993; Prangé et al., 1994; Ducrot et al., 1994) and cebetins (Jalal et al., 1992; Robeson and Jalal, 1993); they have a polycyclic backbone consisting of a highly substituted, partially hydrogenated xanthone system cross-linked by two carbon-carbon covalent bonds to a partially hydrogenated anthraquinone system (Figs 1 and 2). One of the yellow compounds was previously known as CBT (*Cercospora beticola* toxin) (Schlosser,1971), a name still kept by Nasini's group for the major metabolite, very probably identical to cebetin B, reported in Figure 2 (Arnone et al., 1993). Before the recent structural characterization of the yellow toxins, the name CBT had been constantly assigned to the purified phytotoxic extracts of *C.beticola* cultures (all produced by Nasini's group) utilized for biological activity studies.

Investigations of CBT effects on plasma membrane functions, started in 1979, showed in maize roots a marked inhibition of K^+-uptake and of active H^+-extrusion, a very early decrease of the transmembrane electric potential, and a consistent inhibition of the plasma membrane K^+-depend-

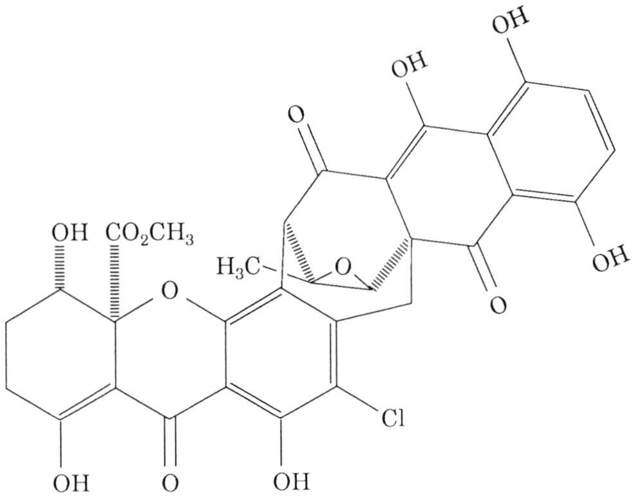

Figure 1. Cebetin A or Beticolin 1.

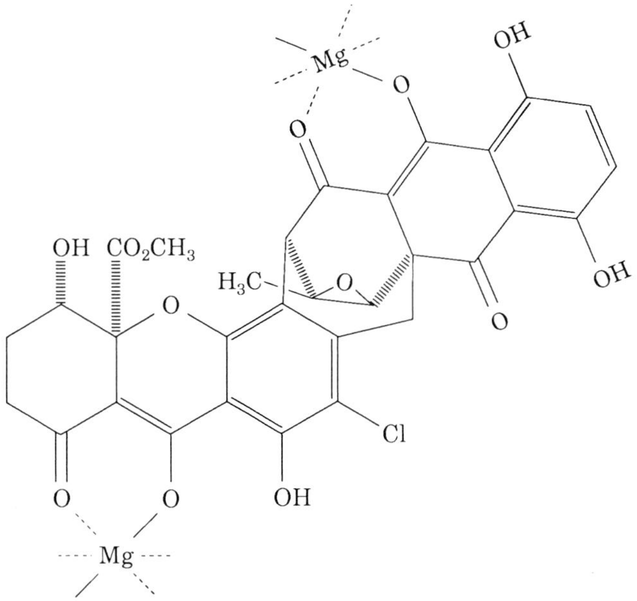

Figure 2. CBT or Cebetin B.

ent ATPase activity (Macrì and Vianello, 1979a, b; Tognoli et al., 1979; Macrì et al., 1980). The possibility that the toxin primarily affects the ATPase(s) associated with microsomal membranes was supported by the inhibitory effect on ATP-dependent proton translocation in pea stem microsomal vesicles (Macrì et al., 1983). Recently, the study of CBT effects on proton translocation has been resumed with membrane preparations containing significantly less contamination by endoplasmic reticulum, mitochondria, or Golgi apparatus, and enriched in the vanadate-sensitive plasma membrane ATPase. It was concluded that the ATP-driven proton transport is 50% inhibited by 2×10^{-6} M CBT on 5 min incubation, an effect which increases with incubation time in consequence of the inhibition of ATPase itself (Blein et al., 1988; Aducci et al., 1992).

The inhibition of the enzyme has been carefully investigated with the use of purified preparation of H^+-ATPase from maize roots reconstituted into artificial proteoliposomes (Aducci et al., 1992). It has been established that the amount of CBT necessary to inhibit 50% of the ATPase activity steadily decreases with the incubation length, and that the inhibition increases on increasing pH. The maximal velocities of H^+ transport and of ATP hydrolysis are independent of toxin concentration. Lineweaver-Burk plot analysis of inhibition data as a function of ATP concentration suggests a competition for Mg-ATP between enzyme and toxin (Rossignol et al., 1989).

Similar results have been obtained in experiments with beticolin-1 and reconstituted ATPase (Simon-Plas et al., 1995), thus confirming that the plasma membrane H^+-ATPase is a direct target for that toxin. It has been recently reported that also the tonoplast ATPase is competitively inhibited by the toxin, indeed with a sensitivity higher than the plasmalemma enzyme (Milat et al., 1992a).

An accurate fluorimetric analysis of beticolins-1 and-2 interaction with biological membranes (Mikes et al., 1994a, b) has shown that in liposomes and plasma membranes the toxins are present in dissociated forms, that the addition of Mg^{2+} to liposomes markedly increases their affinity for a low polarity phase, and that the concentration of their Mg^{2+}-complexes increases at high pH. The results show that the prevailing form of *C.beticola* yellow toxins in the experiments directed to investigate their mode of action is an uncharged complex with Mg^{2+}.

In conclusion, it appears that the perception step consists in the direct interaction of a Mg^{2+}-complex with the plasma membrane, possibly in its headgroup region. The binding to a membrane protein, implicit in the early statement that CBT inhibits the interaction of fusicoccin with its binding sites (Tognoli et al., 1979), has been disproved (Aducci et al., 1992). The transduction step leading to H^+-ATPase inhibition is unclear at present; we can only say that it is accelerated by a high pH which consistently increases the concentration of the active molecule in the membrane.

Brefeldin

Brefeldin A is a macrolide antibiotic (Fig. 3) with high antiviral and cytotoxic activities (Tamura et al., 1968). It is a metabolite of a wide range of fungi (Stoessl, 1981), named differently according to the species name of the producing organism. Decumbin (Singleton et al., 1958) and cyanein (Betina et al., 1962), the names used in earliest publications, have been superseded by brefeldin A (BFA), first introduced in 1963 (Härri et al., 1963).

Toxicity of BFA on plants was an early finding. The toxin is able to interfere with a number of different physiological processes such as germination (Singleton et al., 1958), root growth (Betina et al., 1963) and mitosis (Betina and Murin, 1964). More recently, its possible role in safflower stem and head blight caused by *Alternaria carthami* has been discussed (Tietjen et al., 1983a), as well as its importance in suppression of phytoalexin biosynthesis in safflower suspension cultures (Tietjen et al., 1983b; Tietjen and Matern, 1984). Safflower leaf tissues and protoplasts were not damaged by BFA; also the cell membrane potential was unaffected by the toxin (Tietjen et al., 1983a). Thus, the plasma membrane was excluded as a primary site of action of BFA. In consideration of its hydrophobic character the toxin might easily get across a membrane and interact with some cytosolic component.

A wholly new approach to the mode of action of BFA began in 1985 when it was demonstrated that the toxin causes intracellular accumulation of high mannose-type G protein and inhibition of its expression on cell surface in baby hamster kidney cells infected with vesicular stomatitis virus (Takatsuki and Tamura, 1985). A similar investigation with a primary culture of rat hepatocytes showed that the toxin very efficiently blocks protein translocation from endoplasmic reticulum to the Golgi complex (Misumi et al., 1986). The inhibition of the secretory pathway is accompanied by the disassembly of the Golgi apparatus and the redistribution of some Golgi proteins into the endoplasmic reticulum (Fujiwara et al., 1988; Lippincott-Schwartz et al., 1989). These changes

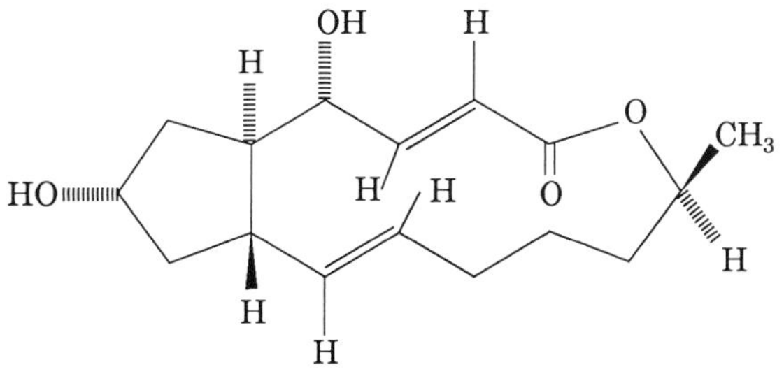

Figure 3. Brefeldin A.

are induced by very low doses of the toxin (2×10^{-7} M) and are fully reversed when the product is removed (Klausner et al., 1992). Detoxification in mammalian cells is probably induced by formation of conjugates arising by a glutathione-S-transferase activity (Brüning et al., 1992).

In the course of a few years, brefeldin A has become a formidable tool for studying the molecular bases of the mechanism involved in intracellular trafficking; the literature dealing with its effects on this process and the consequent applications in the study of complicated cell biology problems has increased tremendously during the last decade (Klausner et al., 1992). Recent investigations have shown that the effects of BFA on membrane traffic may be attributed to its ability to prevent binding of cytosolic coat proteins onto membranes. Such process is known to be controlled by membrane-bound trimeric G-proteins and BFA can prevent the assembly of non-clathrin coatomer complex onto Golgi membranes (Donaldson et al., 1990; Orci et al., 1991; Robinson and Kreis, 1992; Narula et al., 1992) by inhibiting the GTP-dependent interaction of ADP-ribosylation factor (ARF) with Golgi membrane (Donaldson et al., 1992). It has been proposed that BFA may inhibit either the activation of the trimeric G-proteins or their coupling with the effector.

The discovery of the remarkable effects of BFA on cellular processes occurring within the endomembrane system in animal tissues has made obvious a reassessement of the mode of action of this toxin also in plants, with particular attention to its influence on the structure and function of the plant Golgi apparatus. In plants BFA has been shown to affect processing of N-linked glycans and to inhibit, as in animals, protein secretion. The latter effect has been observed in sycamore maple cells (Driouich et al., 1993) and more recently in maize, where BFA was shown to inhibit the secretion of the auxin-binding protein ABP1 (Jones and Herman, 1993). BFA also reduces wall deposition in red algae (Garbary and Phillips, 1993) and inhibits transport of soluble proteins from the trans-Golgi to the vacuole (Holwerda et al., 1992; Gomez and Chrispeels, 1993). However, while protein transport of secretory proteins is affected, protein synthesis is not (Driouich et al., 1993). The initial response of plant cells to BFA is a clustering of the Golgi stack, (Driouich et al., 1993; Satiat-Jeunemaitre and Hawes, 1992a, 1993) not observed in animal cells, which leads to its complete disintegration (Satiat-Jeunemaitre and Hawes, 1992a, b, 1993; Rutten and Knuiman, 1993). This effect is reversible but occurs at concentrations much higher than those active in animal cells (Satiat-Jeunemaitre and Hawes, 1992a). However, in plants there is no evidence for a redistribution of disassembled Golgi membrane protein markers into the ER (Satiat-Jeunemaitre and Hawes, 1992a; Henderson et al., 1994); ER proteins can be rearranged by BFA but their redistribution always follows the Golgi apparatus disruption and is a spatially separate and distinct process (Henderson et al., 1994). Thus, as in animal systems, BFA is a powerful tool to study the organisation and function of the plant Golgi stack (Satiat-Jeunemaitre and Hawes, 1994).

Fusicoccin

Discovery and early studies

Fusicoccum (Phomopsis) amygdali, the causative agent of peach and almond canker (Graniti, 1962), produces a major phytotoxic metabolite named fusicoccin (FC). This compound was iso-lated, together with a number of related metabolites, for the first time by Ballio et al. (1964) from stirred cultures of the fungus. It was demonstrated that FC is responsible for systemic symptoms such as wilt of leaves occurring during pathogen infection (Graniti, 1964; Turner and Graniti, 1976) and that this is due to the ability of the compound to induce stomata opening which leads to uncontrolled transpiration (Turner and Graniti, 1969).

The structure of FC (Fig. 4) was elucidated by Ballio et al. (1968a) and independently by Barrow et al. (1968; 1971). It was established that FC is the α–glucoside of a carbotricyclic

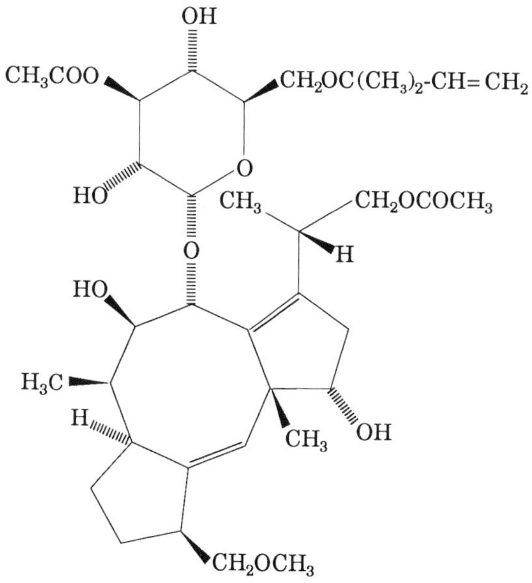

Figure 4. Fusicoccin.

diterpene whose basic ring skeleton is found in other natural products (Muromtsev et al., 1994).

When consistent amounts of the toxin were obtained from cultures of the fungus scaled up at the pilot plant level (Ballio et al., 1968b) extensive investigations on the mechanism of its toxicity became possible. Early studies revealed that FC induces a marked increase of growth by cell enlargement in every higher plant species tested and in a large variety of organs (for a review see Graniti et al., 1994). This response is accompanied by acidification of the incubation medium, an effect strikingly similar to that induced by the natural growth hormone auxin. Systematic investigations on the biological properties of FC have ascertained that the toxin interferes with a number of physiological and biochemical processes of higher plants, such as breaking of seed dormancy, solute transport, regulation of intracellular and extracellular pH, control of transmembrane electrical potential (for a review see Graniti et al., 1994), thus reinforcing the interest in the study of its mechanism of action. Further investigations have established that FC-induced acid secretion is associated with increased K^+ uptake and transmembrane potential hyperpolarization (for a review see Graniti et al., 1994). From these circumstantial pieces of evidence, Marrè (1979) proposed that the rationale for FC action was the stimulation of a, at that time, presumptive plasma membrane electrogenic proton pump. According to this model, the increased proton extrusion triggers cell wall loosening and therefore cell enlargement; on the other hand K^+ uptake by the guard cells contributes to the increase of turgor leading to stomata opening and therefore to toxicity. The increased proton gradient across the plasmalemma and the consequent hyperpolarization of the membrane potential, builds on the driving force for solute uptake.

The synthesis of a tritiated FC derivative with biological activities comparable to those of the parent compound allowed to start studies on FC interaction with plant cells. FC binding sites were detected for the first time in the microsomal fraction of corn coleoptiles (Dorhmann et al., 1977). By further studies it was established that the binding sites are widespread in the plasma membrane of plants from liverworts to Angiosperms, while they are absent in prokariotes, algae, fungi and animal cells (Meyer et al., 1993). Interestingly, the toxicity of FC to higher plants (Chain et al., 1971) parallels the presence of FC receptors (Meyer et al., 1993). More recently, it has been shown that *Arabidopsis* mutants with reduced sensitivity to FC also exhibit reduced FC binding capability (Holländer-Czytko and Weiler, 1994).

Studies over the years, carried out by different groups on different species, have resulted in a detailed characterization of the biochemical properties of FC receptors, such as temperature and pH dependence, kinetics of the binding reaction, saturability and specificity towards FC and a number of derivatives and analogues (for a review see Aducci et al., 1995). These systematic investigations have led to the identification of the structural requirements necessary for receptor recognition (Ballio et al., 1981). Scatchard analysis of binding data have allowed to estimate affinity constants and concentration of FC receptors in plant tissues (for a review see Aducci et al., 1995).

The very restricted variability of all so far characterized properties of FC receptors suggests a highly conserved function in the plasma membrane of higher plants, identifiable with the general control of transport processes achieved by regulating the activity of the H^+-ATPase.

Endogenous ligands

The above-mentioned constitutive role of FC receptors in plant cells seems to be in contrast to the very rare opportunity of plants to encounter FC under natural conditions. As far as we know, this is restricted to almond and peach trees infected by *F. amygdali*. Hence, to solve this apparent paradox, the occurrence in plant tissues of endogenous ligands for the FC perception system, which could act as the physiological regulators of the H^+-ATPase, has been envisaged. Evidence for their existence in higher plants has been reported since 1980 (Aducci et al., 1980a) by showing that substances contained in crude extracts of maize and spinach competed with labeled FC in a binding test to FC receptors. An immunoaffinity phase, obtained by immobilizing anti-FC antibodies onto a Sepharose matrix (Marra et al., 1988) was also used for partial purification of these compounds. Since then, despite similar efforts carried out by another group, who claimed to have partially purified from plant extracts substances inducing stomata opening in the dark (Weiler et al., 1990), research on the identification of these bioactive endogenous plant metabolites has not progressed any further. A rather different point of view has been presented by a Russian team headed by G.S. Muromtsev. In 1980, they reported on the occurrence of endogenous FCs in maize cobs (Muromtsev et al., 1980), but a reassessment of their investigation by Aducci et al. (1985) with the help of a very sensitive RIA for FC detection could not confirm their conclusion. Nevertheless, the occurrence of compounds strictly related to FC in different plant species was claimed by these authors in later work relied on more refined techniques such as GC/MS (Muromtsev et al., 1989) and radioimmunoassay for the identification of the endogenous FCs (Babakov et al., 1994).

The occurrence of endogenous ligands or of endogenous FCs is therefore still an open question. A clear answer to it is nevertheless crucial for the definition of the physiological function of FC receptors, i.e., whether their function is the perception of a novel class of naturally occurring plant regulators or unrelated to FC-binding ability.

Fusicoccin receptors: Purification and identification

Purification of FC receptors has for a long time been hampered by their very scarce concentration in plant tissues and by their lability, likely due to the action of endogenous hydrolases (Aducci et al., 1984). Nevertheless, early studies clarified that FC receptors are distinct from the H^+-ATPase and gave a rough evaluation of their molecular size, which resulted (by gel filtration) in the range 70–90 kDa (Pesci et al., 1979). In recent years progress in the purification of plasma membranes and in protein chromatography has allowed a thorough purification of FC receptors and, consequently, the start of sequence studies.

Octylglucoside solubilisation and affinity chromatography by FC-agarose (De Boer et al., 1989) and by biotin-FC (Korthout et al., 1994), resulted in the purification from oat membranes of 30 and 31 kDa proteins (SDS-PAGE) which in the latter case were still able to bind tritium-labeled FC. FPLC of detergent-solubilised membranes from *Vicia faba* leaves (Feyerabend and Weiler, 1989) or *Commelina communis* leaves (Oecking and Weiler, 1991) yielded a major protein migrating at 34 kDa on SDS-PAGE or two proteins at 30.5 a and 31.6 kDa, respectively. In photoaffinity experiments performed by the same authors, the same polypeptides resulted after being labeled by a radioactive photoactivatable azido-derivative of FC. Solubilisation of FC receptors from maize shoots plasma membranes by a detergent-free method (Aducci et al., 1984) and their purification by HPLC (Aducci et al., 1993) afforded on SDS-PAGE two major proteins of 30 and 90 kDa. The larger was identified as the functional FC receptor on the basis of photoaffinity labeling experiments which showed that, differently from the 30 kDa protein, it became tagged under UV irradiation, not impairing the binding ability of FC receptors (Aducci et al., 1993).

Fusicoccin receptors: Primary structure of the 30 kDa proteins

The above-mentioned purification procedures yielded amounts of pure 30 kDa sufficient for partial sequencing, since the protein has a blocked N-terminus. Sequencing was carried out after internal cleavage and purification of the resulting peptides. Primary structure information was obtained independently by three groups (Marra et al., 1994; Korthout and de Boer, 1994; Oecking et al., 1994). Their results agreed in demonstrating that the 30 kDa proteins recovered in purified fractions of FC receptors belong to the family of 14-3-3 proteins. Members of this class have been isolated first from mammalian brain (Moore and Perez, 1967) and later from a wide range of eukaryotes (Aitken et al., 1992). They are dimeric under native conditions, share the same 30 kDa molecular mass for the two monomers and a pI range of 4.5–5.0; furthermore, they are present as multiple isoforms which usually differ from each other for a few substitutions. After initial uncertainties about their physiological role an increasing number of regulatory functions has been attributed to them, such as tryptophan and tyrosine hydroxylases activation (Ichimura et al., 1987, 1988), PKC inhibition (Toker et al., 1990), association with DNA binding complexes (Lu et al., 1992; De Vetten et al., 1992). In general, they appear as central regulators of eukariotic signaling pathways, likely acting by kinase modulation or protein-protein interactions. Hence, the 30 kDa protein associated with the FC perception system would be, up to now, the unique 14-3-3 protein with characteristics of a plasma membrane receptor for extracellular signals. This stimulatory situation raises two main questions, one related to the intracellular location of FC receptors, the other to the high structural conservation of 14-3-3 proteins. In fact, while different pieces of evidence support the location of FC receptors at the apoplastic face of plasma membrane (Aducci et al., 1980b; Rubinstein, 1982; Feyerabend and Weiler, 1988), members of the 14-3-3 family do not seem to possess features typical of membrane-spanning proteins. Moreover, in a recent paper the location of FC receptors at the cytoplasmic face of plasma membrane is favoured (de Boer and Korthout, 1996). This remains a crucial point to be dealt with. 14-3-3 proteins are widespread also in FC-insensitive organisms where they share multiple regulatory functions; hence, in view of the above-mentioned very high homology of the primary structure of 14-3-3 members, FC-binding ability should have evolved from very limited amino acid variations. In the light of these discrepancies, a FC-receptorial role for the 30 kDa protein cannot be considered conclusive. In this respect, the above reported identification as the FC receptor of a 90 kDa protein accompanying the 30 kDa one in maize tissues must be taken into consideration. Further work is necessary to clarify this still unsettled point; it is necessary on one hand to identify and functionally express in heterologous systems the 30 kDa isoform(s) involved in FC perception or transduction, and on the other to elucidate the primary structure of the 90 kDa protein from maize.

Fusicoccin mode of action

The original proposal that all the physiological and pathological effects induced by FC are dependent on the stimulation of the plasma membrane H^+-ATPase has found experimental confirmation in studies performed by different groups on *in vivo* and *in vitro* systems (Marrè, 1979; Beffagna et al., 1977; Rasi-Caldogno and Pugliarello, 1985; Rasi-Caldogno et al., 1986; Blum et al., 1988; Schulz et al., 1990; De Michelis et al., 1991). Important results have been obtained from *in vitro* experiments performed on crude and, more recently, on purified plasmalemma vesicles (Rasi-Caldogno et al., 1993; Johansson et al., 1993; Lanfermeijer et al., 1994), but the stimulation of proton transport activity was also obtained in proteoliposomes where FC receptors and H^+-

ATPase from maize had been simultaneously incorporated (Aducci et al., 1988; Marra et al., 1992). These studies have firmly established the role of FC receptors as the primary target of FC and as obligatory components for its enzyme activation function. Instead, the very mechanism of enzyme activation by FC is still debated with different hypotheses. Some of them are based on indirect evidence, and, up to now, a satisfactory understanding of the molecular mechanism of plant H^+-ATPase regulation by different effectors is still lacking.

It is now generally accepted that the C-terminal region of P-type H^+-ATPases is an autoinhibitory domain sensitive to several effectors, such as light, lysophospholipids, fungal elicitors, FC, which have the capacity to displace it and, in consequence, to activate the enzyme. The mechanism appears to be widespread in plants and fungi, despite the limited sequence homology of their H^+-ATPases. This hypothesis has been recently reinforced by the evidence that the H^+-ATPase from *Neurospora crassa*, a fungus usually insensitive to FC, can be activated by FC when it is reconstituted into liposomes with plant FC receptors (Marra et al., 1995). A reasonable conclusion from these results is that in organisms as fungi or yeasts, the H^+-ATPases are functionally related to plants and can share with them the same type of regulation. The insensitivity of the above organisms to FC is therefore due to the lack of a FC perception system and probably appeared later in evolution. In this respect, FC represents now not only a tool for studying the mechanism of action of some phytohormones like auxin, but also a probe to investigate the more general problem of the regulation of P-type H^+-ATPase, an enzyme which plays a very important role in a number of processes in plants controlled by different effectors.

The recent evidence produced on the mechanism of FC-induced H^+-ATPase activation has led to the proposal of two different hypotheses: the direct interaction of the FC-receptor complex with the H^+-ATPase or the modification of the H^+-ATPase by specific enzymes (i.e., phosphorylation-dephosphorylation).

Lately, we have obtained some results with a solubilised FC-activated H^+-ATPase from maize roots which show that FC brings about a permanent modification of the enzyme affecting predominantly its H^+-pumping activity (Marra et al., 1996). The purified solubilised enzyme retains in fact the FC activation even in the absence of FC binding activity and of 14-3-3 like proteins, reputed to be FC receptors, thus suggesting that the activated state of H^+-ATPase is the consequence of a covalent modification resistant to detergent treatment and chromatographic fractionation. This hypothesis is sustained by the proved *in vitro* phosphorylation of oat root H^+-ATPase by a membrane-bound protein kinase (Schaller and Sussmann, 1988). However, the evaluation by different authors of the effect of phosphorylation on the enzyme activity has yielded contrasting results. In some cases the stimulation of H^+-ATPase appears to be the consequence of a dephosphorylation reaction (Kauss et al., 1992; Vera-Estrella et al., 1994), in others phos-

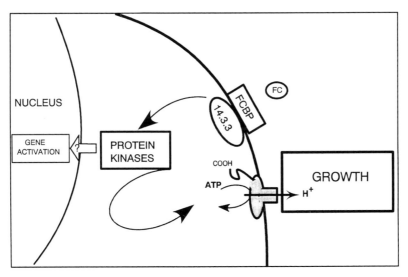

Figure 5. Hypothetical model of fusicoccin mode of action.

phorylation appears instead to be essential for activation (Sekler et al., 1994; Suzuki et al., 1992). Whatever the real mechanism, the evidence that 14-3-3 proteins are part of FC signaling machinery strengthens the point that a kinase-mediated reaction controls the H^+-ATPase activity and opens new perspectives in the study of the transduction system leading to H^+-ATPase activation. A hypothetical model of FC signal transduction pathway is reported on Figure 5. Once more, FC may be considered a tool in plant physiology; the elucidation of its transduction pathway will help to clarify at the molecular level the regulation of H^+-ATPase, a target common to different physiological signals.

Prehelminthosphorol

The sesquiterpene prehelminthosphorol (Fig. 6) has been recently shown to represent the main phytotoxic metabolite of *Bipolaris sorokiniana* (syn. *Helminthosporium sativum*, teleomorph *Cochliobolus sativus*), a fungus pathogenic to barley and wheat (Carlson et al., 1991). It had been previously isolated from culture filtrates of *C. sativus* (De Mayo et al., 1965; Nukina et al., 1975; Aldridge and Turner, 1970) and *C. setariae* (Nukina et al., 1975; Nukina and Marumo, 1976), and from a species of *Bipolaris* pathogenic of Johnson grass (Pena-Rodriguez et al., 1988). Previous studies have attributed to the structurally-related terpenes helminthosphorol (Tamura et

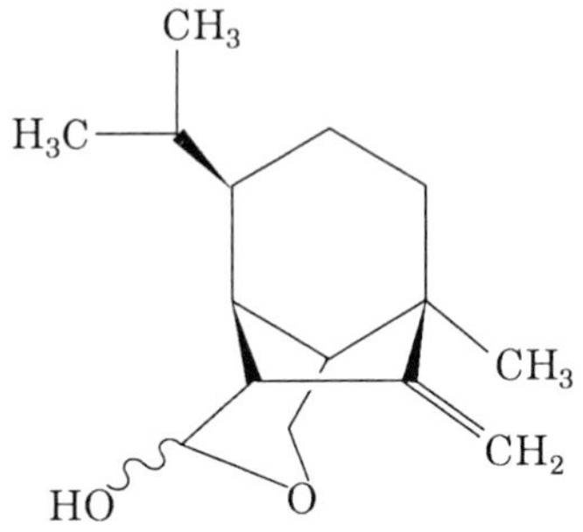

Figure 6. Prehelminthosporol.

al., 1965; De Mayo et al., 1961) and helminthosphoral a prominent role in the induction of disease symptoms. Subsequently, it was demonstrated (De Mayo et al., 1965) that helminthosphoral is an artefact of the isolation procedure arising from an ethyl substituted derivative of prehelminthosphorol only formed when the extraction of the metabolites was carried out in the presence of ethanol. Recently, prehelminthosphorol has been proposed to play a major role in the early stages of infection of barley by *B. sorokiniana* (Nilsson et al., 1993). Prehelminthosphorol and its 9-hydroxy-derivative, also a metabolite of the fungus, show antiviral activity *in vitro* (Aldridge and Turner, 1970).

The interference of prehelminthosphorol with the membranes of root cells has been proposed by Liljeroth et al. (1994), and very recently supported by an investigation reporting its interaction with three enzymes located at the plasma membrane (Olbe et al., 1995). It was observed that prehelminthosporol is slightly inhibitory on the phosphohydrolytic activity of barley roots plasma membrane vesicles made leaky by Triton X-100, is strongly inhibitory on proton pumping into the same vesicles (no Triton added), and fairly inhibitory on ATP-dependent Ca^{2+} uptake. The H^{+}- and the Ca^{2+} pumps are differently sensitive to high concentrations of the toxin, suggesting different toxin-induced membrane permeabilities for these ions. Perhaps prehelminthosporol acts as a protonophore, thus quite differently from fusicoccin (and very likely, also differently from the peptides of *Rhynchosporium secalis*) which affects the plasma membrane H^{+}-ATPase only after binding to a specific membrane receptor (Aducci et al., 1995).

Rather, the action of prehelminthosporol on the H^{+}-ATPase is reminiscent of that reported for *Pseudomonas syringae* pv. *syringae* toxins which disrupt membrane permeability and inhibit the phosphohydrolytic activity of inside-out vesicles (Che et al., 1992; Di Giorgio et al., 1994; Camoni et al., 1995). Except for some minor differences, the effect of prehelminthosporol on 1,3-β-glucan synthase is very similar to that on the H^{+}-ATPase (Olbe et al., 1995).

In conclusion, the main target of prehelminthosporol is probably the plasma membrane and the H$^+$-ATPase located on it. Evidence for a receptor perceiving the toxin, or for a transduction step is not available.

Rhynchosporium secalis toxins

Rhynchosporium secalis is the causal agent of leaf scald disease in barley and several other grasses. The involvement of phytotoxins in the development of symptoms was suggested in early studies (Ayesu-Offei and Clare, 1971). Host-selective low molecular weight toxins, called rhynchosporosides (Auriol et al., 1978; Mazars and Auriol, 1983), were produced in cultures of *R.secalis* isolated from naturally infected barley in Montana, USA, or in France. More recently, a toxic host-nonselective glycoprotein, endowed of elicitor-like activity (Mazars et al., 1990), was isolated from culture filtrates of the fungus (Mazars et al., 1984; Mazars et al., 1987; Mazars et al., 1989a, b). Crude membrane preparations of barley were shown to bind the rhynchosporosides (Pio Beltran and Strobel, 1978; Mazars et al., 1983); glucose and cellobiose competed with the toxins, thus indicating a lectin-like nature of the binding sites. Some results concerning the structure of these toxins have been considered questionable (Daly and Knoche, 1982; Kono et al., 1981).

In the course of studies on the resistance inheritance of barley to the pathogen, neither rhynchosporosides nor glycoprotein toxins were detected in necrosis-inducing culture filtrates of *R.secalis* race US238.1 (virulent on cv Atlas but avirulent on the near-isogenic cv Atlas 46) (Wevelsiep et al., 1991). Instead, three phytotoxic peptides with Mr lower than 10 kDa were isolated (NIP1, NIP2, and NIP3) and purified to electrophoretic homogeneity (Wevelsiep et al., 1991). Their possible involvement in the development of disease symptoms is supported by several lines of evidence (Wevelsiep et al., 1991).

NIP1 and NIP3 stimulate the phosphohydrolytic activity of the H$^+$-ATPase of barley plasma membrane vesicles in a dose-dependent manner (Wevelsiep et al., 1993). In spite of the same toxicity *in vivo*, NIP2 is inactive in the ATPase assay. Stimulation of the hydrolytic activity is lost when the enzyme is solubilized, suggesting the requirement for binding sites. The occurrence of a binding protein of about 65 kDa in barley plasma membrane vesicles is indicated by photoaffinity labeling and affinity chromatography experiments with NIP3 (Wevelsiep et al., 1993). These results evidence a similarity between the mode of action of fusicoccin and of NIP1 and NIP3. In order to evaluate how close this similaritiy is, it is necessary to produce more data on the covalent structure and conformation of NIPs, to purify and characterize their binding protein(s), to investigate the possible competition between NIPs and fusicoccin in their interaction with the binding sites.

Interestingly, NIP1 is also a race-specific elicitor of defense responses in barley cultivars carrying the resistance gene *Rrs1*(Hahn et al., 1993). Thus, the signal carried by this toxic metabolite triggers in the plant cell two different processes, one of offence to, the other of defence by the host. The elicitor inactivity of NIP3, which on the other hand stimulates the H^+-ATPase, demonstrates that the two processes proceed, at least in part, independently.

Zinniol

Zinniol (Fig. 7) was isolated for the first time from cultivars of *Alternaria zinniae*, the fungus responsible for leaf spot and seedling blight of zinnia, sunflower, and marigold (White and Starrat, 1967; Starrat, 1968). It is also produced by some other plant pathogenic species of *Alternaria* (Stoessl et al., 1979; Barash et al., 1981; Cotty et al., 1983; Cotty and Misaghi, 1984; Stierle et al., 1993) and by *Phoma macdonaldii* (Sugawara and Strobel, 1986), and has been found in sunflower tissues infected with *P.macdonaldii* (Sugawara and Strobel, 1986). The preparation of ^{3}H-labeled zinniol has afforded a ligand suitable for studying its interaction with the plant cell (Thuleau et al., 1988). It binds reversibly and in a saturable manner to intact protoplasts and microsomal preparations from zinniol-sensitive cell lines. The affinity for the ligand is lower in the case of resistant cell lines. Perception of zinniol by apparently specific receptors at the plasma membrane results in the stimulation of Ca^{2+} uptake (Thuleau et al., 1988). The influx of $^{45}Ca^{2+}$ into protoplasts of sensitive carrot lines is concentration-dependent, with a maximum at 10^{-6} M zinniol; a further increase of concentration brings about a decrease of the

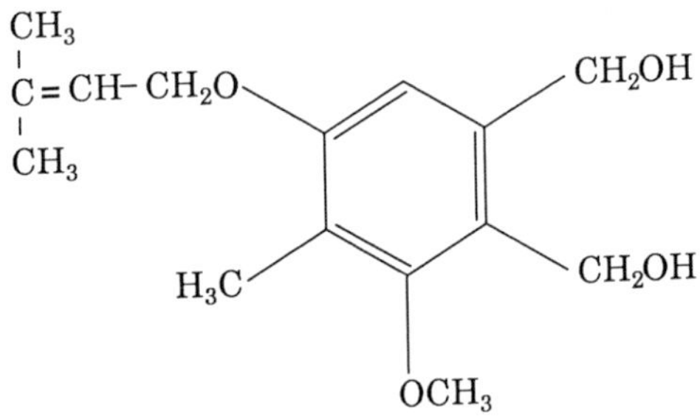

Figure 7. Zinniol.

effect, which is abolished at 10^{-4} M concentration. In protoplasts of resistant cell lines the influence of zinniol on calcium uptake is much less pronounced. From these results it was deduced that the activation properties of the toxin are directly linked to the interaction with the binding sites. Calcium-channel blockers of the verapamil type, besides inhibiting Ca^{2+} uptake into protoplasts, bind to plant membrane preparations; their binding in sensitive lines is 50% inhibited by zinniol, while that in resistant lines is unaffected (Thuleau et al., 1988). The enhancement by zinniol of Ca^{2+} uptake in sensitive protoplasts is contrasted in a dose-dependent manner by the calcium-channel blocker bepridil. Consequently, zinniol can be considered an agonist for a particular class of plant calcium-channels. It is well known that the intracellular level of calcium influences a number of molecular processes in living organisms; thus, a zinniol-promoted abnormal increase of intracellular calcium consequent to the infection of a zinniol-producing pathogen might seriously affect the metabolic equilibrium of the host and hence cause an irreversible damage to it.

Conclusions

Phytotoxins belong to the large number of molecules carrying signals from the environment to plants. Their effects on the plant cell may result from: a) the direct action on a plasma membrane physiological target; b) the binding to a specific recognition system on the plasma membrane; c) the interaction with an intracellular target. Examples of the three types of interaction have been presented in this discussion. Interestingly, in the first case the toxin can escape from a perception-transduction pathway: in fact, both CBT and pre-helminthosporol affect membrane transport by changing membrane permeability and by acting directly on H^+-ATPase. Instead, the two other types of interaction are consistent with the presence of a perception-transduction system, even if evidence so far available is not sufficient to produce a detailed picture of the pathway. The best characterized toxins belonging to case b) are FC and zinniol. The FC perception step has been characterized in detail and specific receptors have been extensively investigated; so far, attempts to elucidate the pathway of signal transduction to the plasma membrane H^+-ATPase have yielded inconclusive results. On the contrary, limited data are available for zinniol receptors, but clear evidence has been provided that the binding to protoplasts induces an enhancement of calcium influx, suggestive of a cascade system analogous to that observed in animals and in plants for other signal molecules.

The third type of interaction involving an intracellular target applies to BFA. The extensive study of the biological properties of this compound has led to the discovery of remarkable effects on the structure of Golgi apparatus and on protein secretion, not only in plants, but also in animal cells. In recent years, BFA has become a frequently used tool in cell biology and has significantly contributed to the comprehension of the complex problem of intracellular trafficking, both in animals and in plants. Besides FC, BFA represents a good example of the potentially wider applications originating from the study of the molecular mechanisms involved in perception and transduction of signals carried by phytotoxins.

References

Aducci, P., Ballio, A., Crosetti, G. and Federico, R. (1980a) Fusicoccin receptors. Evidence for endogenous ligand. *Planta* 148: 208–210.

Aducci, P., Federico, R. and Ballio, A. (1980b) Interaction of a high molecular weight derivative of fusicoccin with plant membranes. *Phytopathol. Mediter.* 19: 187–188.

Aducci, P., Ballio, A., Fiorucci, L. and Simonetti, E. (1984) Inactivation of solubilised fusicoccin binding sites by endogenous plant hydrolases. *Planta* 160: 422–427.

Aducci, P., Ballio, A., Evidente, A., Federico, R., Iasiello, I., Marra, M. and Randazzo, G. (1985) On the reported occurrence of a fusicoccin conjugate in maize cobs. *Phytochemistry* 24: 1097–1099.

Aducci, P., Ballio, A., Blein, J.-P., Fullone, M.R., Rossignol, M. and Scalla, R. (1988) Functional reconstitution of a proton-translocating system responsive to fusicoccin. *Proc. Natl. Acad. Sci. USA* 85: 7849–7851.

Aducci, P., Cozzella, M. L., Di Giorgio, D., Fogliano, V. and Marra, M. (1992) *Cercospora beticola* toxin: a reassessment of some *in vitro* effects. *Plant Sci.* 84: 53–57.

Aducci, P., Ballio, A., Fogliano, V., Fullone, M.R., Marra, M. and Proietti, N. (1993) Purification and photoaffinity labeling of fusicoccin receptors from maize. *Europ. J. Biochem.* 214: 339–345.

Aducci, P., Marra, M., Fogliano, V. and Fullone, M.R. (1995) Fusicoccin receptors: perception and transduction of the fusicoccin signal. *J. Exp. Bot.* 46: 1463–1478.

Aitken, A., Collinge, D.B., van Heusden, B.P.H, Isobe, T., Roseboom P.H., Rosenfeld, G. and Sol, J. (1992) 14-3-3 proteins: a highly conserved, widespread family of eucaryotic proteins. *Trends Biochem. Sci.* 17: 498–501.

Aldridge, D.C. and Turner, V.B. (1970) 9-hydroxy-prehelminthosphorol, a metabolite of *Cochliobolus (Helminthosporium) sativus*. *J. Chem. Soc.* (C): 686–688.

Arnone, A., Nasini, G., Merlini, L., Ragg, E. and Assante, G. (1993) Secondary mould metabolites. Part 41. Structure and biosynthesis of *Cercospora beticola* toxin (CBT). *J. Chem. Soc., Perkin Trans.* 1: 145–151.

Assante, G., Locci, R., Camarda, L., Merlini, L. and Nasini, G. (1977) Screening of the genus *Cercospora* for secondary metabolites. *Phytochemistry* 16: 243–246.

Auriol, P., Strobel, G., Pio Beltran, J. and Gray, G. (1978) Rynchosporoside, a host-selective toxin produced by *Rhynchosporium secalis.*, the causal agent of scald disease of barley. *Proc. Natl. Acad. Sci. USA* 75: 4339–4343.

Ayesu-Offei, E.N. and Clare, B.G. (1971) Symptoms of scald disease induced by toxic metabolites of *Rhynchosporium secalis*. *Austr. J. Biol. Sci* 24: 169–174.

Babakov, A.V., Bartova, L.M., Galina, U., Oganian, R.R., Voblikova, V.D., Maysurian, A.N., Dridze, I.L. and Muromtsev, G.S. (1994) Endogenous fusicoccin-like ligand revealed in higher plants by radioreceptor and radioimmunoassays. *FEBS Lett.* 351: 243–245.

Ballio, A., Chain, E B., De Leo, P., Erlanger, B.F., Mauri, M. and Tonolo, A. (1964) Fusicoccin: a new wilting toxin produced by *Fusicoccum amygdali* Del. *Nature* 203: 296.

Ballio, A., Brufani, M., Casinovi, C.G., Cerrini, S., Fedeli, W., Pellicciari R., Santurbano, B. and Vaciago, A. (1968a) The structure of fusicoccin. *Experientia* 24: 631–635.

Ballio, A., Carilli, A., Santurbano, B. and Tuttobello, L. (1968b). Produzione di fusicoccina in scala pilota. *Annali dell' Istituto Superiore di Sanità* 4: 317–332.

Ballio, A., Federico, R. and Scalorbi, D. (1981) Fusicoccin structure-activity relationships: *In vitro* binding to microsomal preparations of maize coleoptiles *Physiol. Plant.* 52: 476–481.

Barash, I., Mor, H., Netzer, D. and Kashman, Y. (1981) Production of zinniol by *Althernaria carthami* and its phytotoxic effect on carrot. *Physiol. Plant Pathol.* 19: 7–16.

Barrow, K.D., Barton, D.H.R., Chain, E.B., Ohnsorge, U.V.F. and Thomas R. (1968) The constitution of fusicoccin. *Chem. Comm.* 1198–1200.

Barrow, K.D., Barton, D.H.R., Chain, E.B., Ohnsorge, U.V.F. and Thomas, R. (1971) The constitution of fusicoccin. *J. Chem. Soc.* (C) 1265–1273.

Beffagna, N., Cocucci, S. and Marrè, E. (1977) Stimulating effect of fusicoccin on K-activated ATPase in plasmalemma preparations of higher plant tissue. *Plant Sci. Lett.* 8: 91–98.

Betina, V. and Murin, A. (1964) Inhibition of mitotic activity in root tips of *Vicia faba* by the antibiotic cyanein. *Cytologia (Tokyo)* 29: 370–374.

Betina, V., Nemec, P., Dobias, J. and Barath, Z. (1962) Cyanein, a new antibiotic from *Penicillium cyaneum*. *Folia Microbiol. Prague* 7: 353–357.

Betina, V., Nemec, P. and Barath, Z. (1963) Growth inhibition of *Allium cepa* roots by the antibiotic cyanein. *Naturwissenschaften* 50: 696.

Blein, J.-P., Bourdil, I., Rossignol, M. and Scallà, R. (1988) *Cercospora beticola* toxin inhibits vanadate-sensitive H+ transport in corn root membrane vesicles. *Plant Physiol.* 88: 429–434.

Blum, W., Key, G. and Weiler, E.W. (1988) ATPase activity in plasmalemma-rich vesicles isolated by aqueous two-phase partitioning from *Vicia faba* mesophyll and epidermis: characterization and influence of abscisic acid and fusicoccin. *Physiol. Plant.* 72: 297–287.

Brüning, A., Ishikawa, T., Kneusel, R.E., Matern, U., Lottspeich, F. and Wieland, F.T. (1992) Brefeldin A binds to glutathione S-transferase and is secreted as glutathione and cysteine conjugates by chinese hamster ovary cells. *J. Biol. Chem.* 267(11): 7726–7732.

Camoni, L., Di Giorgio, D., Marra, M., Aducci, P. and Ballio A. (1995) *Pseudomonas syringae* pv. *syringae* phytotoxin reversibly inhibit the the plasma membrane H+-ATPase and disrupt unilamellar liposomes. *Biochem. Biophys. Res. Comm.* 214: 118–124.

Carlson, H., Nilsson, P., Jansson, H.B. and Odham, G. (1991) Characterization and determination of prehelminthosphorol, a toxin from the plant pathogenic fungus *Bipolaris sorokiniana*, using liquid chromatography/mass spectrometry. *J. Microbiol. Meth.* 3: 259–269.

Chain, E.B., Mantle, P.G. and Milborrow, B.V. (1971) Further investigations on the toxicity of fusicoccins. *Physiol. Plant Pathol.* 1: 495–514.

Che, F.S., Kasamo, K, Fukuchi, N., Isogai, A. and Suzuki, A. (1992) Bacterial phytotoxins, syringomycin, syringostatin and syringotoxin exert their effect on the plasma membrane H+-ATPase partly by a detergent-like action and partly by inhibition of the enzyme. *Physiol. Plant.* 86: 818–824.

Cotty, P.J. and Misaghi, I.J. (1984) Zinniol production by *Alternaria* species. *Phytopathology* 74: 785–788.

Cotty, P.J., Misaghi, I.J. and Hine, R.B. (1983) Production of zinniol by *Alternaria tagetica* and its phytotoxic effect on *Tagetes erecta*. *Phytopathology* 73: 1326–1328.

Daly, J.M. and Knoche, H.W. (1982) The chemistry and biology of pathotoxins exibiting host-selectivity. *In:* D.S. Ingram and P.H. Williams (eds): *Advances in Plant Pathology*, Vol. 1. Academic Press, New York, pp 83–138.

Daub, M.E. and Ehrenshaft, M. (1993) The photoactivated toxin cercosporin as a tool in fungal photobiology. *Physiol. Plant.* 89: 227–236.

de Boer, A.H. and Korthout, H. (1996) 14-3-3 homologues play a central role in fusicoccin signal transduction pathway. *J. Plant Growth Regul.* 18: 99–105.

de Boer, A.H, Watson, B.A and Cleland, R.E. (1989) Purification and identification of the fusicoccin binding protein from oat root plasma membrane. *Plant Physiol.* 89: 250–259.

De Mayo, P., Spencer, E.Y. and White, R.W. (1961) Helminthosporal the toxin from *Helminthosporium sativum* I. Isolation and characterization. *Can. J. Chem.* 39: 1608–1612.

De Mayo, P., Williams, R.E. and Spencer, E.Y. (1965) Terpenoids VIII, the immediate precursors of helminthosphoral and helminthosphorol. *Can. J. Chem.* 43: 1357–1365.

De Michelis, M.I., Rasi-Caldogno, F., Pugliarello, M.C. and Olivari, C. (1991) Fusicoccin binding to its plasma membrane receptor and activation of the plasma membrane H+-ATPase. II. Stimulation of H+-ATPase in a plasma membrane fraction purified by phase-partitioning. *Bot. Acta* 104: 265–271.

De Vetten, N.C., Lu, G. and Ferl, R.G. (1992) A corn protein associated with the G-box binding complex has homology to brain regulatory proteins. *Plant Cell* 4: 1295–1307.

Di Giorgio, D., Camoni, L. and Ballio, A. (1994) Toxins of *Pseudomonas syringae* pv *syringae* affect H+-transport across the plasma membrane of maize. *Physiol. Plant.* 91: 741–746.

Dohrmann, U., Hertel R., Pesci, P., Cocucci, S.M., Marrè, E., Randazzo, G. and Ballio A. (1977) Localization of *in vitro* binding of fungal toxin fusicoccin to plasma-membrane-rich fractions from corn coleoptiles. *Plant Sci. Lett.* 9: 291–299.

Donaldson, J.G., Lippincott-Schwartz, J., Bloom, G.S., Kreis, T.E. and Klausner R.D. (1990) Dissociation of a 110-kD peripheral membrane protein from the Golgi apparatus is an early event in brefeldin A action. *J. Cell Biol.* 111: 2295–2306.

Donaldson, J.G., Finazzi, D. and Klausner, R.D. (1992) Brefeldin A inhibits Golgi membrane-catalysed exchange of guanine nucleotide onto ARF protein. *Nature* 360: 350–352.

Driouich, A., Zhang, G.F. and Staehelin, A.L. (1993) Effect of Brefeldin A on the structure of the Golgi apparatus and on the synthesis and secretion of proteins and polysaccharides in sycamore maple *(Acer pseudoplatanus)* suspension-cultured cells. *Plant Physiol.* 101: 1336–1343.

Ducrot, P.-H., Milat, M.-L., Blein, J.-P. and Lallemand, J.-Y. (1994) The yellow toxins produced by *Cercospora beticola*. Revised structures of beticolin 1 and 3. *J. Chem. Soc., Chem. Comm.* 2215–2216.

Feyerabend, M and Weiler, EW. (1988) Characterization and localization of fusicoccin binding sites in leaf tissues probed with a novel radioligand. *Planta* 174: 115–122.

Feyerabend, M. and Weiler, E.W. (1989) Photoaffinity labeling and partial purification of the putative plant receptor for the fungal wilt-inducing toxin, fusicoccin. *Planta* 178: 282–290.

Fujiwara, T., Oda, K., Yokota, S., Takatsuki, A. and Ikehara, Y. (1988) Brefeldin A causes disassembly of the Golgi complex and accumulation of secretory proteins in the endoplasmic reticulum. *J. Biol. Chem.* 263: 18545–18552.

Garbary, D.J. and Phillips, D. (1993) Cell wall deposition in the red alga *Ceramium strictum* – Are microtubules involved? *Am. J. Bot.* 80(suppl): 24.

Gomez, L. and Chrispeels, M.J. (1993) Tonoplast and soluble vacuolar proteins are targeted by different mechanisms. *Plant Cell* 5: 1113–1124.

Graniti, A. (1962) Azione fitotossica di *Fusicoccum amygali* Del. su mandorlo (*Prunus amygdali* St.). *Phytopathol. Mediter.* 1: 182–185.

Graniti, A. (1964) Qualche dato di fitotossicità della 'fusicoccina A', una tossina prodotta *in vitro* da *Fusicoccum amygdali. Phytopathol. Mediter.* 3: 125–128.

Graniti, A., Ballio, A. and Marrè, E. (1994) *Fusicoccum (Phomopsis) amygdali. In*: U.S Sing, K. Kohmoto and R.P. Singh (eds): *Pathogenesis and Host Specificity in Plant Diseases*, Vol. 2., Elsevier, Oxford, pp 103–117.

Hahn M., Jüngling S. and Knogge, W. (1993) Cultivar-specific elicitation of barley defense reactions to the phytotoxic peptide NIP1 from *Rhynchosporium secalis. Mol. Plant Microbe Interact.* 6: 745–754.

Härri, E., Loeffler, W., Sigg, H.P., Stähelin, H. and Tamm, C. (1963) Über die Isolierung neuer Stoffwechselprodukte aus *Penicillium brefeldianum* Dodge. *Helvet. Chim. Acta.* 46: 1235–1243.

Henderson, J., Satiat-Jeunemaitre, B., Napier, R. and Hawes, C. (1994) Brefeldin A-induced disassembly of the Golgi apparatus is followed by disruption of the endoplasmic reticulum in plant cells. *J. Exp. Bot.* 45(279): 1347–1351.

Höllander-Czytko, H. and Weiler, E.W. (1994) Isolation and biochemical characterization of fusicoccin-insensitive variant lines of *Arabidopsis thaliana* (L.) Heynh. *Planta* 195: 188–194.

Holwerda, B.C., Padgett, H.S. and Rogers, J.C. (1992) Proaleurain vacuolar targeting is mediated by short contiguous peptide interactions. *Plant Cell* 4: 307–318.

Ichimura, T., Isobe, T., Okuyama, T., Yamauchi, T. and Fujisawa, H. (1987) Brain 14-3-3 protein is an activator protein that activates tryptophan 5-monooxygenase and tyrosine 3-monooxigenase in the presence of Ca^{2+}, calmodulin-dependent protein kinase II. *FEBS Lett.* 219: 79–82.

Ichimura, T., Isobe, T., Okuyama, T., Takahashi, N., Araki K, Kuwano, R. and Takahashi, Y. (1988) Molecular cloning of cDNA coding for brain-specific 14-3-3 protein, a protein kinase-dependent activator of tyrosine and tryptophan hydroxylases. *Proc. Natl. Acad. Sci. USA* 85: 7084–7088.

Jalal, M.A.F., Hossain, M.B., Robeson, D.J. and van der Helm, D (1992) *Cercospora beticola* phytotoxins: cebetins that are photoactive, Mg^{2+}-binding, chlorinated anthraquinone-xanthone conjugates. *J. Am. Chem. Soc.* 114: 5967–5971.

Johansson, F., Sommarin, M. and Larsson, C. (1993) Fusicoccin activates the plasma membrane $H^+ATPase$ by a mechanism involving the C-terminal inhibitory domain. *Plant Cell* 5: 321–327.

Jones, A.M. and Herman E.M. (1993) KDEL-containing auxin-binding protein is secreted to the plasma membrane and cell wall. *Plant Physiol.* 101: 595–606.

Kauss, H., Jeblick, W. and Conrath, U. (1992) Protein kinase inhibitor K-252a and fusicoccin induce similar initial changes in ion transport of parsley suspension cells. *Physiol. Plant.* 85: 483–488.

Klausner, R.D., Donaldson, J.D. and Lippincott-Schwartz, J. (1992) Brefeldin A: insights into the control of membrane traffic and organelle structure. *J. Cell Biol.* 116(5): 1071–1080.

Kono, Y., Knoche, H.W. and Daly, J.M. (1981) Structure: fungal host-specific. *In*: R.D. Durbin (ed.): *Toxins in Plant Disease*, Academic Press, New York, pp 221–257.

Korthout, H. and de Boer AH. (1994) A fusicoccin binding protein belongs to the family of 14-3-3 brain protein homologs. *Plant Cell* 6: 1681–1692.

Korthout, H., Van der Hoeven, P., Wagner, M.J., Van Hunnik, E. and de Boer A.H. (1994) Purification of the fusicoccin-binding protein from oat root plasma membrane by affinity chromatography with biotinylated fusicoccin. *Plant Physiol.* 105: 1281–1288.

Lanfermeijer, F.C. and Prins, H.B.A. (1994) Modulation of $H^+ATPase$ activity by fusicoccin in plasma membrane vesicles from oat *Avena sativa* L. roots. *Plant Physiol.* 104: 1277–1285.

Liljeroth, E., Franzon-Almgren, I. and Gustavsson, M. (1994) Effect of prehelminthosphorol a phytotoxin produced by *Bipolaris sorokiniana* on barley roots. *Can. J. Bot.* 72: 558–563.

Lippincott-Schwartz, J., Yuan, L.C., Bonifacino, J.S. and Klausner R.D. (1989) Rapid redistribution of Golgi proteins into the ER in cells treated with brefeldin A: evidence for membrane cycling from Golgi to ER. *Cell* 56: 807–813.

Lu, G., De Lisle, A., de Vetten, N.C. and Ferl, R. (1992) Brain protein in plants: an *Arabidopsis* homolog to neurotransmitter pathway activators is part of a DNA binding complex. *Proc. Natl. Acad. Sci. USA* 89: 11490–11494.

Macrì, F. and Vianello, A. (1979a) Photodynamic activity of cercosporin on plant tissues. *Plant Cell Environ.* 2: 267–271.

Macrì, F. and Vianello, A. (1979b) Inhibition of K⁺ uptake, H⁺extrusion and K⁺activated ATPase, and depolarization of transmembrane potential in plant tissue treated with *Cercospora beticola* toxin. *Physiol. Plant Pathol.* 15: 161–170.

Macrì, F., Vianello, A., Cerana, R. and Rasi-Caldogno, F. (1980) Effects of *Cercospora beticola* toxin on ATP level of maize roots and on the phosphorylating activity of isolated pea mitochondria. *Plant Sci. Lett.* 18: 207–214.

Macrì, F., Dell' Antone, P. and Vianello, A. (1983) ATP-dependent proton uptake inhibited by *Cercospora beticola* toxin in pea stem microsomal vesicles. *Plant Cell Environ.* 6: 555–558.

Marra, M., Aducci, P. and Ballio, A. (1988) Immunoaffinity chromatography of fusicoccin. *J. Chromatogr.* 440: 47–51.

Marra, M., Ballio, A., Fullone, M.R. and Aducci, P. (1992) Some properties of a functional reconstituted plasmalemma H⁺-ATPase activated by fusicoccin. *Plant Physiol.* 98: 1029–1034.

Marra, M., Fullone, M.R., Fogliano, V., Masi, S., Mattei, M., Pen, J. and Aducci, P. (1994) The 30 kD protein present in purified fusicoccin receptor preparation is a 14-3-3-like protein. *Plant Physiol.* 106: 1497–1501.

Marra, M., Ballio, A., Battirossi, P., Fogliano, V., Fullone, M.R., Slayman, C.L. and Aducci, P. (1995) The fungal H⁺-ATPase from *Neurospora crassa* reconstituted with the fusicoccin receptors senses fusicoccin signal. *Proc. Natl. Acad. Sci. USA* 92: 1599–1603.

Marra, M., Fogliano, V., Zambardi, A., Fullone, M.R., Nasta, D. and Aducci, P. (1996) The H⁺-ATPase purified from maize root plasma membranes retains fusicoccin *in vivo* activation. *FEBS Lett.* 382: 293–296.

Marrè, E. (1979) Fusicoccin: a tool in plant physiology. *Annu. Rev. Plant Physiol.* 30: 273–288.

Mazars, C. and Auriol, P. (1983) Production *in vitro* et *in vivo* de rhynchosporosides par *Rhynchosporium secalis*. *Comp. Rend. Acad. Sci.* 296: 681–683.

Mazars, C., Auriol, P., Rafenomananjava, D. (1983) Rhynchosporosides binding by barley proteins. *Phytopathol. Z.* 107: 1–8.

Mazars, C., Hapner, K.D. and Strobel, G.A. (1984) Isolation and partial characterization of a phytotoxic glycoprotein from culture filtrates of *Rhynchosporium secalis*. *Experientia* 40: 1244–1247.

Mazars, C., Rossignol, M. and Auriol, P. (1987) Purification par HPLC de glycoproteines phytotoxiques produites par *Rhynchosporium secalis* (Oud) Davis. *Bio-Sci.* 6: 63–68.

Mazars, C., Rossignol, M., Marquet, P.Y. and Auriol, P. (1989a) Reassessment of toxic glycoprotein isolated from *Rhynchosporium secalis* (Oud) Davis culture filtrates. *Plant Sci.* 62: 165–174.

Mazars, C., Poletti, P., Petitprez, M., Albertini, L. and Auriol, P. (1989b) Plugging of the xylem vessel of barley induced by a high molecular weight phytotoxic glycoprotein from *Rhynchosporium secalis*. *Can. J. Bot.* 67: 2077–2084.

Mazars, C., Lafitte, C., Marquet, P.Y., Rossignol, M. and Auriol, P. (1990) Elicitor-like activity of the toxic glycoprotein isolated from *Rhynchosporium secalis* (Oud) Davis culture filtrates. *Plant Sci.* 69: 11–17.

Meyer, C., Waldkötter, R., Sprenger, A., Schlösser, U.G., Lether, M. and Weiler, E.W. (1993) Survey of the taxonomic and tissue distribution of microsomal binding sites for the non-host selective fungal phytotoxin, fusicoccin. *Z. Naturforsch.* 48C: 595–602.

Mikes, V., Milat, M.-L., Pugin, A. and Blein J.-P. (1994a) *Cercospora beticola* toxins. VII. Fluorometric study of their interactions with biological membranes. *Biochim. Biophys. Acta* 1195: 124–130.

Mikes, V., Lavernet, S., Milat, M.-L., Collange, E., Paris, M. and Blein J.-P. (1994b) *Cercospora beticola* toxins. Part VI: preliminary studies of protonation and complexation equilibria. *Biophys. Chem.* 52: 259–265.

Milat, M.-L., Fraichard, A., Blein, J.-P. and Pugin, A. (1992a) Effects of beticolin, the yellow *Cercospora beticola* toxin on the plasmalemma ATPase activity and on the tonoplast ATPase and pyrophosphatase activities. *In: Abstracts from the Ninth International Workshop on Plant Membrane Biology*, Monterey, California, July 19–24, 1992, p. 222.

Milat, M.-L., Prangé, T., Ducrot, P.-H., Tabet, J.-C., Einhorn, J., Blein, J.-P. and Lallemand J.-Y. (1992b). Structures of the beticolins, the yellow toxins produced by *Cercospora beticola*. *J. Am. Chem. Soc.* 114: 1478–1479.

Milat, M.-L., Blein, J.-P., Einhorn, J., Tabet, J.-C., Ducrot, P.-H. and Lallemand, J.-Y. (1993) The yellow toxins produced by *Cercospora beticola*. Part II: isolation and structure of beticolins 3 and 4. *Tetrahedron Lett.* 34: 1483–1486.

Misumi, Y., Misumi, Y., Miki, K., Taratswuki, A., Tamura, G. and Ikehara, Y. (1986) Novel blockade by Brefeldin A of intracellular transport of secretory proteins in cultured rat hepatocytes. *J. Biol. Chem.* 261: 11398–11403.

Moore, P.V. and Perez, V.J. (1967) Physiological and biochemical aspects of nervous integration. *In:* F.D. Carlson (ed.): *Symposium on Physiological and Biochemical Aspects of Nervous Integration.* Prentice Hall, Woods Hole, pp 343–359.

Muromtsev, G.S., Kobrina, N.S., Voblikova, V.D. and Koreneva, V.M. (1980) Fusicoccin-like substance in maize ears. *Izv. Akad. Nauk. SSSR. Biol.* 2: 897–902.

Muromtsev, G.S., Voblikova, V.D., Kobrina, N.S., Koreneva, V.M., Sadovskaya, V.L. and Stolpakova, V.V. (1989) Fusicoccin in higher plants. *Biochem. Physiol. Pflanzen* 185: 261–268.

Muromtsev, G.S., Voblikova, V.D., Kobrina, N.S., Koreneva, V.M., Krasnopolskaya, L.M. and Sadovskaya, V.L. (1994) Occurrence of fusicoccanes in plants and fungi. *J. Plant Growth Regul.* 13: 39–49.

Narula, N., McMorrow, I., Plopper, G., Doherty, J., Matlin, K., Burke, B. and Stow, J. (1992) Identification of a 200 kD, Brefeldin-sensitive protein on Golgi membranes. *J. Cell Biol.* 117: 27–38.

Nilsson, P., Akesson, H., Jansson, H.B. and Odham, G. (1993) Production and release of the phytotoxin prehelminthosphorol by *Bipolaris sorokiniana* during growth. *FEMS Microbiol. Ecol.* 102: 91–98.

Nukina, M. and Marumo, S. (1976) Aversion factors, antibiotics among different strains of a fungal species aversion factors of *Cochliobolus setarie. Agricult. Biol. Chem.* 40: 2121–2123.

Nukina, M., Hattori, H. and Marumo, S. (1975) *cis*-Sativenediol, a plant growth promoter produced by fungi. *J. Am. Chem. Soc.* 97: 2542–2543.

Oecking, C. and Weiler, E.W. (1991) Characterization and purification of the fusicoccin binding complex from plasma membrane of *Commelina communis. Europ. J. Biochem.* 199: 685–689.

Oecking, C., Eckerskorn, C. and Weiler E.W. (1994) The fusicoccin receptor of plants is a member of the 14-3-3 superfamily of eukaryotic regulatory proteins. *FEBS Lett.* 352: 163–166.

Olbe, M., Sommarin, M., Gustavsson, M. and Lundborg, T. (1995) Effect of the fungal pathogen *Bipolaris sorokiniana* toxin prehelminthosphorol on barley roots plasma membrane vesicles. *Plant Pathol.* 44: 625–635.

Orci, L., Tagaya, M., Amherdt, M., Perrelet, A., Donaldson, J.D., Lippincott-Schwartz, J., Klausner, R.D. and Rothman, J.E. (1991) Brefeldin A, a drug that blocks secretion, prevents the assembly of non-clathrin-coated buds on Golgi cisternae. *Cell* 64: 1183–1195.

Pena-Rodriguez, L.M. Armingeon, N.A. and Chilton, W.S. (1988) Toxins from weed pathogens. I. Phytotoxins from a *Bipolaris* pathogen of Johnson grass. *J. Nat. Prod.* 51: 821–828.

Pesci, P., Tognoli, L., Beffagna, M. and Marrè, E. (1979) Solubilization and partial purification of a fusicoccin-receptor complex from maize microsomes. *Plant Sci. Lett.* 15: 313–322.

Pio Beltran, J. and Strobel, G. (1978) Rhynchosporoside binding proteins of barley. *FEBS Lett.* 96: 34–36.

Prangé, T., Neuman, A., Milat, M.-L., Blein, J.-P. (1994) The yellow toxin produced by *Cercospora beticola*. Structure of beticolins 2 and 4. *Acta Cristallogr.* B51: 308–314.

Rasi-Caldogno, F. and Pugliarello, M.C. (1985) Fusicoccin stimulates the H⁺ATPase of plasmalemma in isolated membrane vesicles from radish. *Biochem. Biophys. Res. Comm.* 133: 280–285.

Rasi-Caldogno, F., De Michelis, M.I., Pugliarello, M.C. and Marrè, E. (1986) H⁺-pumping driven by the plasma membrane ATPase in membrane vesicles from radish: stimulation by fusicoccin. *Plant Physiol.* 82: 121–125.

Rasi-Caldogno, F., Pugliarello, M.C., Olivari, C. and De Michelis, M.I. (1993) Controlled proteolysis mimicks the effect of fusicoccin on the plasma membrane H⁺ATPase. *Plant Physiol.* 103: 391–396.

Robeson, D.J. and Jalal, M.A.F. (1993) A *Cercospora* isolate from soybean roots produces cebetin B and cercosporin. *Phytochemistry* 33: 1546–1548.

Robinson, M.S. and Kreis, T.E. (1992) Recruitment of coat protein onto Golgi membranes in intact and permeabilized cells: Effects of Brefeldin A and G protein activators. *Cell* 69: 129–138.

Rossignol, M., Bourdil, I., Santoni, V. and Blein, J.-P. (1989) Interaction between plasma membrane H⁺ ATPase and phytotoxins: use of reconstituted systems. *In*: J. Dainty, M.I. De Michelis, E. Marrè and F. Rasi-Caldogno (eds): *Plant Membrane Transport: the Current Position*. Elsevier, Amsterdam, pp 379–384.

Rubinstein, B. (1982) Regulation of H⁺ excretion. Role of protein released by osmotic shock. *Plant Physiol.* 69: 945–49.

Rutten, T.L.M. and Knuiman, B. (1993) Brefeldin A effects on tobacco pollen tubes. *Europ. J. Cell Biol.* 69: 129–138.

Satiat-Jeunemaitre, B. and Hawes, C. (1992a) Redistribution of a Golgi glycoprotein in plant cells treated with Brefeldin A. *J. Cell Sci.* 103: 1153–1166.

Satiat-Jeunemaitre, B. and Hawes, C. (1992b) Reversible dissociation of the plant Golgi apparatus by Brefeldin A. *Biol. Cellul.* 74: 325–328.

Satiat-Jeunemaitre, B. and Hawes, C. (1993) The distribution of secretory products in plant cells is affected by Brefeldin A. *Cell Biol Int.* 17: 183–193.

Satiat-Jeunemaitre, B. and Hawes, C. (1994) G.A.T.T. (A general agreement on traffic and transport) and Brefeldin A in plant cells. *Plant Cell* 6: 463–467.

Schaller, G.E. and Sussmann, M.R. (1988) Phosphorylation of the plasma membrane H⁺ATPase of oat roots by a calcium-stimulated protein kinase. *Planta* 173: 509–518.

Scheffer, R.P. and Pringle, R.B. (1964) Uptake of Helminthosporium victoriae toxin by oat tissue. *Phytopathology* 54: 832–835.

Schulz, S., Oelgemöller, E. and Weiler, E.W. (1990) Fusicoccin action in cell suspension culture of *Corydalis sempervirens. Planta* 183: 83–91.

Schlosser, E. (1971) The *Cercospora beticola* toxin. *Phytopathol. Mediter.* 10: 154–158.

Sekler, I., Weiss, M. and Pick, U. (1994) Activation of the *Dunaliella acidophila* plasma membrane H⁺ATPase by trypsin cleavage of a fragment that contain a phosphorylation site. *Plant Physiol.* 105: 1125–1132.

Simon-Plas, F., Gomès, E., Milat, M.-L. and Blein J.-P. (1995) Some aspects of the inhibition of maize roots plasmalemma H⁺-ATPase by beticolin-1. *In: Abstracts from the 10th International Workshop on Plant Membrane Biology, Regensburg*, Germany, August 6–11, 1995, P 11.

Singleton, V.L., Bohonos, N. and Ullstrup, A.J. (1958) Decumbin, a new compound from a species of *Penicillium. Nature* 181: 1072–1073.

Starratt, A.N. (1968) Zinniol: a major metabolite of *Alternaria zinniae. Can. J. Chem.* 46: 767–770.

Stierle, A., Hershenhorn, J. and Strobel, G. (1993) Zinniol-related phytotoxins from *Alternaria cichorii*. *Phytochemistry* 32: 1145–1149.

Stoessl, A. (1981) Structure and biogenetic relations: fungal nonhost-specific. *In*: R. D. Durbin (ed.): *Toxins in Plant Disease*, Academic Press Inc., New York, pp 110–219.

Stoessl, C.H., Unwin, J.B. and Stothers, J.B. (1979) Metabolites of *Alternaria solani*. Part 5. Biosynthesis of altersolanol A and incorporation of altersolanol A-^{13}C into altersolanol B and macrosporin. *Tetrahedron Lett.* 2481–2484.

Sugawara, F. and Strobel G. (1986) Zinniol, a phytotoxin, is produced by *Phoma macdonaldii*. *Plant Sci.* 43: 19–23.

Suzuki, Y.S., Wang, Y. and Takemoto, J.Y. (1992) Syringomycin-stimulated phosphorilation of the plasma membrane H$^+$ATPase from red beet storage tissue. *Plant Physiol.* 99: 1314–1320.

Takatsuki, A. and Tamura, G. (1985) Brefeldin A, a specific inhibitor of intracellular translocation of vesicular stomatitis virus G protein: intracellular accumulation of high-mannose type G protein and inhibition of its cell surface expression. *Agricult. Biol. Chem.* 49: 899–902.

Tamura, S., Sakurai, A., Kainuma, K. and Takai, M. (1965) Isolation of helminthosphorol as a natural plant growth regulator and its chemical structure. *Agricult. Biol. Chem.* 29: 216–221.

Tamura, G., Ando, K., Suzuki, S., Takatsuki, A. and Arima, K. (1968) Antiviral activity of Brefeldin A and Verrucarin A. *J. Antibiot.* 21: 160–161.

Thuleau, P., Graziana, A., Rossignol, M., Kauss, H., Auriol, P., Ranjeva, R. (1988) Binding of the phytotoxin zinniol stimulates the entry of calcium into plant protoplast. *Proc. Natl. Acad. Sci. USA* 85: 5932–5935.

Tietjen, K. and Matern, U. (1984) Induction and suppression of phytoalexin biosynthesis in cultured cells of safflower, *Carthamus tinctorius* L. by metabolites of *Althernaria carthami* Chowdhury. *Arch. Biochem. Biophys.* 229: 136–144.

Tietjen, K.G., Schaller, E. and Matern, U. (1983a) Phytotoxin from *Alternaria carthami* Chowdhury: structural identification and physiological significance. *Physiol. Plant Pathol.* 23: 387–400.

Tietjen, K.G., Hunkler, D. and Matern, U. (1983b) Differential response of cultured parsley cells to elicitors from two non-pathogenic strains of fungi. I. Identification of induced products as coumarin derivatives. *Europ. J. Biochem.* 131: 401–407.

Tognoli, L., Beffagna, N., Pesci, P. and Marrè, E. (1979) On the relationship between ATPase activity and FC binding capacity of crude and partially-purified microsomal preparations from maize coleoptiles. *Plant Sci. Lett.* 16: 1–14.

Toker, A., Ellis, C.A., Sellers, L.A., Amess, B. and Aitken, A. (1990) Protein kinase C inhibitor proteins. *J. Biochem.* 191: 421–429.

Turner, N.C. and Graniti, A. (1969) Fusicoccin, a fungal toxin that opens stomata. *Nature* 223: 1070.

Turner, N.C. and Graniti, A. (1976) Stomatal response of two almond cultivars to fusicoccin. *Physiol. Plant Pathol.* 9: 175–182.

Vera-Estrella, R., Barkla, B.J., Higgins, V.J. and Blumwald, E. (1994) Plant defense response to fungal pathogen. *Plant Physiol.* 104: 209–215.

Weiler, E.W, Meyer, C., Oecking, C., Feyerabend, M. and Mithöfer, A. (1990) The fusicoccin receptor of higher plants *In*: C. Lamb and R. Beachy (eds): *Plant Gene Transfer*, Alan R.Liss, New York, pp 153–164.

Weiss, U., Merlini, L. and Nasini, G. (1987) Natural occurring perylenequinones. *Prog. Chem. Org. Nat. Prod.* 52: 1–71.

Wevelsiep, L., Kogel, K.-H. and Knogge, W. (1991) Purification and characterization of peptides from *Rhynchosporium secalis* inducing necrosis in barley. *Physiol. Molec. Plant Pathol.* 39: 471–482.

Wevelsiep, L., Rupping, E. and Knogge, W. (1993) Stimulation of barley plasmalemma H$^+$-ATPase by phytotoxic peptides from the fungal pathogen *Rhynchosporium secalis*. *Plant Physiol.* 101: 297–301.

White, G.A. and Starratt, A.N. (1967) The production of a phytotoxin substance by *Alternaria zinniae*. *Can. J. Bot.*, 45: 2087–2090.

Signal Transduction in Plants
P. Aducci (ed.)
© 1997 Birkhäuser Verlag Basel/Switzerland

Blue light-activated signal transduction in higher plants

W.R. Briggs[1] and E. Liscum[2]

[1]*Department of Plant Biology, Carnegie Institution of Washington, Stanford, CA 94305, USA*
[2]*Division of Biological Sciences, University of Missouri, Columbia, MO 65211, USA*

Summary. In this review, we discuss recent progress in the understanding of various elements of signal transduction cascades activated by blue light. We consider two responses in particular: inhibition of stem growth and phototropism. Included is a discussion of recent work related to the identification of photoreceptors and their chromophores. We also discuss blue light-modulated regulation of gene expression and ion movements. Physiological investigations are included where they appear relevant.

Introduction

As sessile organisms, plants have evolved the capacity to respond to a wide range of environmental signals, one of the most important of which is light. Indeed, virtually every phase of a plant's development is regulated by its light environment. These collective responses are encompassed by the term photomorphogenesis. Recent reviews by Quail (1991), Millar et al. (1994), Furuya (1993), and Smith (1995) deal with the red/far red photoreversible photoreceptors – the phytochromes – while those by Kaufman (1993), Short and Briggs (1994), and Jenkins et al. (1996) deal with photomorphogenesis regulated by blue light photoreceptor systems. Deng (1994), Chamovitz and Deng (1995), and Chory and Susek (1994) have written general reviews, and a recent comprehensive book edited by Kendrick and Kronenberg (1994) provide a broad and up-to-date coverage of virtually all aspects of photomorphogenesis.

As the above reviews testify, progress in understanding the complexities of photoregulated processes mediated by the phytochrome family of photoreceptors has been rapid and dramatic, particularly since the isolation of mutants of *Arabidopsis* (*Arabidopsis thaliana*) impaired in their capacity to respond to light by inhibition of hypocotyl growth (Koornneef et al., 1980). By contrast, progress in understanding photoregulation of plant processes activated by blue light has until recently lagged far behind the phytochrome work. The two volumes of the *Encyclopedia of Plant Physiology* that were published 12 years ago (Shropshire and Mohr, 1983a, b) are heavily devoted to phytochrome, and only a single chapter title has the words "blue light" in it – and even there, the title reads, "Blue-light effects on phytochrome-mediated responses"! The proceedings of two international meetings on blue light, one in 1979 (Senger, 1980) and one in 1984 (Senger, 1984), and a two-volume set of contributed articles on blue light responses (Senger, 1987) are rich in physiology, phenomenology, and discussions of putative chromophores, but signal transduction is hardly mentioned.

Only in the last few years has significant progress in understanding elements in blue light-activated signal transduction been made at the cellular, biochemical, and molecular levels. The purpose of this article is to summarize this recent work. The emphasis is entirely on blue light-activated processes, and work is discussed only if it appears to the authors to have direct relevance to signal transduction with respect to an identifiable physiological response. We will begin with a general consideration of the use of mutants in photomorphogenesis research, next consider blue light-induced inhibition of stem growth, and then discuss recent progress in understanding events in the signal transduction pathway for phototropism. Sections on regulation of gene expression and ion movement will round out this review.

Photomorphogenic mutants

Ever since Koornneef and co-workers (Koornneef et al., 1980) described the *hy* (long *h*ypocotyl) mutants of *Arabidopsis*, mutants have played an increasingly central role in efforts to elucidate the complexities of photomorphogenesis. This statement is witnessed by the number of reviews on photomorphogenesis with mutant analyses as the focal point (Chory, 1993; Chory and Susek, 1994; Deng, 1994; Koornneef and Kendrick, 1994; Liscum and Hangarter, 1994; Whitelam and Harberd, 1994; Chamovitz and Deng, 1995). A few studies with direct relation to areas discussed in this review are presented below.

Photomorphogenic mutants of *Arabidopsis* have now been identified for two apophytochromes (apophyA and apophyB, see Whitelam and Harberd, 1994), a blue light photoreceptor for hypocotyl growth suppression (see Jenkins et al., 1996), a putative blue light photoreceptor for phototropism (see Liscum and Briggs, 1995), proteins required for the biosynthesis of phytochromobilin (see Whitelam and Harberd, 1994), potential signal-transduction components for hypocotyl growth suppression (see Chory, 1992; Whitelam et al., 1993), and phototropism (Liscum and Briggs, 1996), and regulatory components that potentially mediate the transition from an etiolated (dark) to a de-etiolated (light) growth pattern (see Chory and Susek, 1994; Chamovitz and Deng, 1995). Many of these mutants have been essential for sorting out different pathways for photomorphogenesis (Chory, 1992; Liscum and Hangarter, 1994; Jenkins et al., 1996).

One use of photomorphogenic mutants has been to determine whether similar or completely different pathways operate when closely related responses such as phototropism and inhibition of hypocotyl growth are both induced by the same wavelength regions of the spectrum (UV-A and blue). In this case, mutants clearly establish that these two responses are mediated by different photoreceptors and apparently distinct signal transduction pathways (Liscum et al., 1992; Liscum and Briggs, 1995; E. Liscum, unpublished results). Another use of photomorphogenic mutants has been to distinguish between blue light photoreceptor-specific and phytochrome-specific responses where both are induced by blue light (see Liscum and Hangarter, 1994). Such mutants have also been used to determine where blue light photoreceptor- and phytochrome-activated processes converge to mediate similar responses (see Chory, 1992; Liscum, unpublished results).

A photoreceptor for blue light-dependent inhibition of hypocotyl growth in *Arabidopsis*

Apoprotein

One of the *Arabidopsis* mutants isolated by Koornneef et al. (1980), *hy4-2.23N*, showed normal sensitivity to red and far red light with respect to hypocotyl growth inhibition, but was clearly insensitive to blue light. However, hypocotyl growth inhibition is not the only light response affected in *hy4* mutants. Liscum and Hangarter (1991) noted that under blue light, the cotyledons of *hy4-101*, *hy4-102*, *hy4-103*, and *hy4-104* (originally designated *blu1*, *blu2*, *blu3-1* and *blu3-2*, respectively; E. Liscum and R. Hangarter, personal communication) were considerably smaller than wild type, though other major morphological differences were not detected when the plants were grown under white light. Jackson and Jenkins (1995) have recently shown that, in addition to smaller cotyledons, adult *hy4* mutant plants grown under continuous blue light also had longer petioles and flowering stalks, and larger leaves than wild type.

Ahmad and Cashmore (1993) utilized a T-DNA-tagged *hy4* allele (*hy4-2*) to clone the *HY4* gene and found that the wild-type gene encodes a protein with a predicted molecular mass of 75.8 kDa. The N-terminal portion of the protein encoded by the *HY4* gene has remarkable similarity to microbial DNA photolyase, an enzyme known to mediate pyrimidine dimer repair by a blue light-dependent electron transfer (Sancar, 1994). The HY4 protein was subsequently named CRY1 for cryptochrome (Lin et al., 1995a), a commonly used designation for blue light photoreceptors (Gressel, 1979). The CRY1 protein shows 30% sequence identity with seven DNA photolyases, and over certain regions (such as putative chromophore binding sites), identity as high as 70–80% (Ahmad and Cashmore, 1993).

Chromophores

Photolyases have two chromophores: an invariant FAD, and an accessory chromophore which can either be deazaflavin or a pterin such as methenyltetrahydrofolate (MTHF). Among the photolyases, sequence homologies for the second chromophore binding site are greater between enzymes using the same accessory chromophores than between those utilizing different ones (Sancar, 1994). Ahmad and Cashmore (Ahmad and Cashmore, 1993) predicted on the basis of sequence homology that in addition to probable use of FAD as a chromophore, CRY1 should use deazaflavin as the accessory chromophore.

It has now been demonstrated that CRY1, purified from heterologous expression systems, does utilize FAD as a chromophore and binds it non-covalently (Lin et al., 1995a; Malhotra et al., 1995) and at an apparent 1:1 stoichiometry with apoCRY1 (Lin et al., 1995a). Under anaerobic conditions, the FAD can be photoreduced to a stable flavosemiquinone having significant absorption in the green portion of the visible spectrum. In fact, at predicted cellular redox potentials, a significant portion of the bound FAD would be in the semiquinone form. Malhotra et al. (1995) have recently shown that the fluorescence emission spectrum observed with purified recombinant CRY1 was that to be expected if CRY1 had both FAD and a pterin as chromophores. They confirmed chemically that the second chromophore was a pterin, rather than a deazaflavin as predicted by the DNA sequence (Ahmad and Cashmore, 1993). Specifically, CRY1 expressed in *E. coli* appears to bind MTHF. However, it remains an open question whether CRY1 actually binds this pterin *in planta* or a deazaflavin as suggested by the sequence homology. The Cashmore laboratory has recently sequenced an additional 19 mutant alleles at the *HY4* locus, and report a cluster of mutants to have their sequences altered within the putative flavin binding site (Ahmad, 1995), indicating the importance of FAD binding to the physiological response to blue light.

Holoprotein function

Although higher plants, including *Arabidopsis* (Pang and Hays, 1991), clearly have photolyase activity, CRY1 apparently does not function as a photolyase (Lin et al., 1995a; Malhotra et al., 1995). However, its homology to photolyases suggested that CRY1 may function as a blue light photoreceptor for higher plants (Ahmad and Cashmore, 1993). Furthermore, if CRY1 functions as a blue light photoreceptor, it may do so by a novel light-activated electron-transfer mechanism (Malhotra et al., 1995). Three pieces of evidence indicate that CRY1 does indeed function as a photoreceptor. First, as discussed above, purified CRY1 has associated with it the two prosthetic groups found in the folate-type photolyases, namely FAD and the pterin MTHF. This holoCRY1 is capable of absorbing UV-A, blue, and green light, wavelengths in which apparent *hy4* null mutants exhibit reduced hypocotyl growth inhibition relative to wild type (Koornneef et al., 1980; Ahmad and Cashmore, 1993; Lin et al., 1995b). Second, tobacco plants overexpressing the *Arabidopsis HY4* gene showed a clear increase in sensitivity of hypocotyl growth inhibition to UV-A, blue, and green light (Lin et al., 1995b). Finally, *hy4-101* represents an allele that is deficient in both blue and green photosensitivity compared to wild type, but retains normal UV-A sensitivity (Young et al., 1992), as do the *hy4* alleles described by Ahmad and Cashmore (1993). Such an alteration in spectral sensitivity is most easily explained as a lesion affecting binding or function of a single chromophore where more than one chromophore contributes to holoprotein. A similar argument is used with relation to a potential phototropism photoreceptor (see Liscum and Briggs, 1995, and below). Together, the above studies demonstrate conclusively that CRY1 is the photoreceptor for hypocotyl growth inhibition in *Arabidopsis*.

Two additional photolyase/*HY4*-like genes have been identified in photosynthetic organisms, one in *Sinapis alba* (Batschauer, 1993), and one *in Chlamydomonas reinhardtii* (Small et al., 1995). Neither of these genes appear to be functional photolyases (Malhotra et al., 1995; Small et al., 1995), despite an initial report of limited protection from UV-B light conferred by the *S. alba* gene in a photolyase-deficient *E. coli* mutant (Batschauer, 1993). To date, there is no known physiological role for the proteins encoded by these genes. However, because of their homology to DNA photolyase and CRY1, it is reasonable to hypothesize that they may act as photoreceptors receptors for some blue light-activated process(es) in these organisms.

Signal transduction for phototropism

Biochemical studies

In 1988, Gallagher et al. (1988) noted that light induced a change in the phosphorylation state of a membrane protein in etiolated pea seedlings. Subsequent studies showed that blue light was specifically activating the phosphorylation of a plasma membrane protein at multiple sites, a reaction that could be driven either *in vivo* or *in vitro*. The photosensitivity of the system *in vitro* has allowed extensive biochemical characterization both in pea and maize, and to some extent in *Arabidopsis*. Short and Briggs (1994) have presented a detailed discussion of many of its properties.

Several other properties of this light-driven phosphorylation deserve mention. First, the reaction requires free SH groups, as SH-alkylating and -phenylating agents inhibit the reaction (Rüdiger and Briggs, 1995). The SH groups involved are probably located in a hydrophobic environment, since increasing the hydrophobicity of the inhibitor increases its effectiveness. As inhibition occurs regardless of whether the reagent is added before or after irradiation, the SH groups must at minimum participate in some reaction subsequent to the initial photoexcitation. There is evidence from 2-D electrophoresis of the maize tryptic phosphopeptides generated following different fluences of blue light that phosphorylation may, to a limited extent, be hierarchical (J.M. Palmer and W.R. Briggs, unpublished data). Two of the phosphopeptides become phosphorylated at fluences far below those required for full phosphorylation of the others.

The current working hypothesis is that photoreceptor, kinase, and phosphorylation substrate functions are all encompassed in this single protein. First, permeabilization in Triton X-100 has no detectable effect on the quantum efficiency for light-activated phosphorylation (Short et al., 1993), indicating that if additional proteins are involved, they must be tightly associated with the phosphoprotein. Second, an EMS- (ethylmethane sulfonate) derived mutant of *Arabidopsis*, *nph1-2* (previously designated strain JK224, see *Genetic studies*, below), exhibits reduced sensitivity to blue light but not to green light with respect to first positive phototropism (see Konjevic et al., 1992) and shows much reduced levels of the phosphoprotein (Liscum and Briggs, 1995; P. Oeller, E. Liscum, and W.R. Briggs, unpublished results). As previously discussed with respect to CRY1, it is unlikely that a mutation in a protein other than a photoreceptor could cause a change in the spectral sensitivity for a given response, in this case, phototropism. The hypothesis that there are two separate photoreceptors sensitive in blue and green respectively (Konjevic et al., 1992) is unlikely as four additional *nph1* mutants (see below) lack both blue- and green-light photosensitivity for phototropism (Liscum and Briggs, 1995; E. Liscum, unpublished results). Preliminary evidence suggesting that the phosphorylated protein itself may be a kinase, and that the phosphorylation is therefore some sort of autophosphorylation comes from studies demonstrating that the protein has an ATP-binding site (Palmer et al., 1993; Short et al., 1993), a prerequisite for kinase activity.

Physiological studies

There is also a significant body of correlative physiological evidence that the phosphoprotein is in some way involved in phototropism. This evidence is summarized in Table 1. Again, Short and Briggs (1994) present a more detailed discussion of these studies. The many physiological correlations make a strong case for this involvement, but of course correlative evidence does not provide definitive proof. Indeed, one correlation is lacking in the case of maize: phosphorylation *in vivo* shows a considerably higher fluence threshold than phototropism (Palmer et al., 1993a).

Table 1. Correlations between physiological properties of phototropism and phosphorylation

1. Both phototropism and phosphorylation obey the Bunsen-Roscoe reciprocity law (Short and Briggs, 1990).
2. The plant tissues most sensitive to phototropic induction have the highest levels of phosphorylation (Short and Briggs, 1990; Palmer et al., 1993b).
3. Phosphorylation shows dark-recovery kinetics *in vivo* following saturating irradiation similar to those for phototropism (Short and Briggs, 1990; Palmer et al., 1993b)
4. The light-driven phosphorylation is far faster than the physiological response (Short and Briggs, 1990)
5. The activity spectrum for phosphorylation matches the action spectrum for phototropism (Palmer et al., 1993b).
6. The *in vivo* fluence-response curve for phototropism in pea matches that for phosphorylation (Short and Briggs, 1990; Short et al., 1992).

Genetic studies

As discussed above (see *Biochemical studies*), mutants at a single locus, *NPH1* (mutant designation: *nph* for *non-phototropic hypocotyl*), provide convincing genetic evidence that the phosphoprotein is involved in the phototropic response (see Liscum and Briggs, 1995). Furthermore, these mutants provide compelling evidence that the phosphoprotein may be the photoreceptor for phototropism (Konjevic et al., 1992; Liscum and Briggs, 1995), and that this photoreceptor mediates all of the phototropic responses known for *Arabidopsis* (Liscum and Briggs, 1995)

We have also identified and characterized phototropic mutants at three additional loci: *NPH2*, *NPH3*, and *NPH4*. Mutant alleles at these loci show normal levels of the phosphoprotein, and normal blue light-activated phosphorylation of that protein despite their impaired phototropic responses (Liscum and Briggs, 1995). The *nph2* and *nph3* mutants are impaired only with respect to their phototropic responses while the *nph4* mutants are gravitropically impaired as well (Liscum and Briggs, 1996). Together, physiological and biochemical analysis of these mutants has resulted in the tentative ordering of loci in the phototropic signal perception/transduction/-response pathway as *NPH1* → *NPH2/ NPH3* → *NPH4*: where *NPH1* likely encodes the phosphoprotein/photoreceptor, *NPH2* and *NPH3* likely encode signal carriers acting downstream from the phosphoprotein, but upstream from any reactions in common between phototropism and gravitropism, and *NPH4* presumably encodes a protein involved in the differential growth response since mutants at this locus are impaired in two tropic responses.

Recently, we have isolated mutants (two alleles) at an additional locus, *NPH5*, that have impaired phototropic and gravitropic responses (E. Liscum and W.R. Briggs, unpublished data). These mutants exhibit pleiotropic phenotypes resembling those observed in a subset of *constitutive photomorphogenic* (or *cop*) mutants (see Deng, 1994). Analysis of the tropic responses of *nph5*, as well as *cop2*, *cop3*, and *cop4* mutants suggest that the *nph5* mutants are alleles at the *cop2* locus (E. Liscum and W.R. Briggs, unpublished data). It is presently unclear how the *NPH5* (*COP2*?) gene product may function with respect to phototropic/gravitropic signal transduction.

Despite their impaired and null responses to blue and green light respectively, etiolated *nph2* and *nph4* mutant seedlings exhibit phototropic curvature when exposed to UV-A light (Liscum and Briggs, 1996). This UV-A responsiveness can be phenocopied when seedlings are irradiated either with UV-A or with red light from above (so as not to induce a phototropic response) prior to unilateral irradiation with blue light. In both cases there is a clear enhancement of the phototropic response to blue light by the pretreatment (Liscum and Briggs, 1996). Wild-type seedlings preirradiated with UV-A or red light from above also exhibit increased phototropic curvature in response to unilateral blue light. These data, together with those from an examination of the phytochromobilin-deficient *Arabidopsis* mutant *hy1-100* (Chory et al., 1989; Parks and Quail, 1991; Nagatani et al., 1993) indicated that phytochrome phototransformation from Pr to Pfr is required for the enhancement response (Liscum and Briggs, 1996). Further studies (E. Liscum, unpublished results) indicate that both phytochrome A and B play a role in altering the tropic behavior of etiolated *Arabidopsis* seedlings. They evidently do so in a complex manner with distinct as well as overlapping functions and potential antagonistic relationships. The blue light-induced phototropic curvature observed for de-etiolated *nph2* and *nph4* seedlings (Liscum and Briggs, 1996) is therefore presumably a consequence of phytochrome photoconversion during the white light treatment used to induce de-etiolation.

CRY1 is not involved in the perception of phototropic stimuli

Given that CRY1 clearly acts as a blue/UV-A light photoreceptor in *Arabidopsis*, it was possible that it might be the same as (or a partially redundant homologue of) NPH1 or could mediate phosphorylation of that protein, presumed to be somehow involved in the initiation of signal transduction. Three lines of evidence indicate that this is not the case: First, CRY1 is predicted to be a cytoplasmic protein of M_r 75.8 kDa (Ahmad and Cashmore, 1993), whereas the *Arabidopsis* phosphoprotein is a plasma membrane-associated protein with an M_r near 120 kDa (Reymond et al., 1992). Second, membranes from the *hy4-2.23N* deletion mutant show wild-type levels of the phosphoprotein and its light-dependent phosphorylation (Liscum and Briggs, 1995). Third, *hy4* mutants show normal phototropic responses to blue light despite their lack of blue light-dependent hypocotyl growth inhibition, while *nph1* mutants show normal hypocotyl growth inhibition despite their null phototropic phenotype with respect to phototropism (Liscum et al., 1992; Liscum and Briggs, 1995). In addition to their clearly different physiologies, *hy4* and *nph1* mutants are genetically complementary (Liscum and Briggs, 1995).

Potential chromophore(s) for the phototropism photoreceptor

The chemical nature of the chromophore for the phototropism photoreceptor is currently unknown. Candidate molecules suggested as chromophores for blue light responses in general are flavins and carotenoids (see Briggs and Iino, 1983), and more recently pterins (Galland and Senger, 1988) and retinal (Lorenzi et al., 1994). Quiñones and Zeiger (1994) have presented evidence that the level of the carotenoid zeaxanthin, a component of the xanthophyll cycle (Demmig-Adams, and Adams, 1992), was correlated with phototropic sensitivity under several conditions in maize coleoptiles, and suggested that it might be the chromophore for the photoreceptor for phototropism. Although the correlations presented are statistically meaningful, there are at least three reasons why they are not likely biologically meaningful: First, zeaxanthin is found only in plastids, whereas the photoreceptor for phototropism is apparently in the plasma membrane (Gallagher et al., 1988; Palmer et al., 1993b; Short et al., 1993; Liscum and Briggs, 1995). Second, the action spectrum for phototropism, both first positive (Shropshire and Withrow, 1958) and second positive (Everett and Thimann, 1968), has a distinct action peak in the UV-A where carotenoids do not normally absorb. Third, seedlings lacking detectable carotenoids either through treatment with the herbicide norflurazon or because of a genetic lesion early in isoprenoid synthesis exhibit normal phototropic curvatures and normal blue light-induced phosphorylation (Palmer et al., 1996), supporting earlier preliminary studies both with herbicide-treated (Vierstra and Poff, 1981) and mutant plants (Bandurski and Galston, 1951; Labouriau and Galston, 1955; see Galston, 1959).

As mentioned earlier, *Arabidopsis* exhibits sensitivity to green light for hypocotyl growth suppression, in addition to blue and UV-A light, in contrast to other plant species (Baskin and Iino, 1987). The Cashmore group has presented evidence that this anomalous sensitivity to green light may be because the FAD chromophore exists as a flavosemiquinone (Lin et al., 1995a). Similar anomalous sensitivity of *Arabidopsis* to green light is found in phototropism (Steinitz et al., 1985; Konjevic et al., 1989; Konjevic et al., 1992). Hence, it is possible that the green light-sensitive phototropic response of *Arabidopsis* is a consequence of the photoreceptor for phototropism using FAD or some other flavin in a similar semiquinone redox state (Liscum and Briggs, 1995).

There is actually photophysiological evidence that more than one chromophore may be involved in phototropism. Many years ago, Shropshire and Withrow (Shropshire and Withrow, 1958) found that the slope of the fluence-response curves in the UV-A in log-linear plots were steeper than the slopes for wavelengths in the blue, a phenomenon also found for second positive curvature (Everett and Thimann, 1968), although the overall spectrum had the shape expected of a flavin. It is axiomatic that if a single chromophore is involved in a photoresponse and no screening pigments are involved, only the zero intercept should change from one wavelength to the next as a function of relative absorption by the chromophore, but the slopes on a log-linear plot should remain constant. If like CRY1, the photoreceptor for phototropism had both FAD and a second UV-A-absorbing chromophore such as MTHF, this discrepancy could be explained. Clearly, further studies are need to see whether these suggestions are valid.

Regulation of gene expression

Over 15 years ago, Tobin first showed that phytochrome phototransformation from Pr to Pfr could regulate the steady state level of a nuclear-encoded mRNA, in this case that for the small subunit of ribulose-1,5-carboxylase (Tobin, 1978). Several years passed before the first demonstration that this kind of phytochrome regulation was at the transcriptional level (Silverthorne and Tobin, 1984; Berry-Lowe and Meagher, 1985). Since these earlier studies, a very large literature has developed on phytochrome regulation of nuclear gene expression. This regulation turns out to be extremely complex both at the level of promoter elements involved in light regulation (Batschauer et al., 1994; Terzaghi and Cashmore, 1995), and at the photophysiological level (Thompson et al., 1985; Thompson and White, 1991). The different phytochrome-induced changes in mRNA abundance may be positive or negative; the changes occur over minutes or hours, without a lag period or with a lag period of hours; fluence requirements for threshold and saturation may vary greatly, and escape from far-red reversibility may occur in minutes or hours (Thompson et al., 1985).

Similar studies on blue light-induced changes in mRNA abundance were hampered by the fact that it was difficult to unambiguously separate those mediated by blue light photoreceptors from those phytochrome effects on the same mRNAs. Kaufman and coworkers (Kaufman et al., 1985) solved this problem by growing pea seedlings (*Pisum sativum*) under continuous low fluence-rate red light, and then treating the seedlings with single pulses of blue light (see Briggs and Iino, 1983). It was presumed that the blue light pulse would only trivially disturb the long-term Pr/Pfr photoequilibrium, thus permitting the detection of blue light responses mediated by photoreceptors other than phytochrome. Feinbaum et al. (1991) used a similar regime to demonstrate a blue light-induced increase in the activity of the reporter gene β-glucuronidase (*GUS*) when driven by the light-responsive *Arabidopsis* chalcone synthase (*CHS*) promoter (Feinbaum and Ausubel, 1988).

Studies with pea

Using the above technique for sorting blue light photoreceptor-mediated from phytochrome-mediated response, Kaufman and co-workers (1985) demonstrated for the first time an unambiguous effect of blue light on the abundance of a specific mRNA: blue light dramatically increased the steady-state level of the mRNA for the small subunit of rubisco. In addition to the effect on the rubisco small-subunit mRNA, the abundance of several additional mRNA species, some identified as to protein product and some not, has been shown to be regulated by a blue light photoreceptor(s) (Warpeha et al., 1989). The regulation of the levels of these transcripts could be either positive or negative or both, and the fluence relationships also exhibited complexities, with low fluence (LF) and high fluence (HF) requirements. For example, mRNA representing the chlorophyll *a/b*-binding protein family (*CAB*) showed both LF (increased abundance) and HF (decreased abundance to dark level) responses, whereas two other mRNAs, pEA207 and pEA25 showed only an HF response (decreased abundance).

Further studies (Warpeha and Kaufman, 1990a) revealed more complexities in the system. First, if seedlings are grown in complete darkness, *CAB* showed only an LF response while pEA 207 showed no response at all. Since far red light had no effect in reversing the *CAB* response to blue light while it fully reversed a *CAB* response to red, a specific blue light photoreceptor was implicated. There is clearly some developmental control, as in younger red light-grown seedling, the buds showed a LF response for *CAB* and no response at all for pEA 207 whereas the young leaves showed only a HF response for *CAB* and pEA 207. As a further complexity, the *CAB* response here was strongly positive in contrast to the negative response observed in plants grown for a longer period in continuous red light! Finally, red and blue light can act either additively (*CAB*) or antagonistically (pEA25 and pEA46), or blue light can have an effect where red light has none (pEA207).

As with most of the phytochrome-mediated changes in mRNA abundance, the blue light-induced regulation of mRNA abundance was shown to be transcriptional (Marrs and Kaufman, 1989), a conclusion based on results of nuclear run-on transcription experiments. The unexpected complexities in blue light regulation of mRNA abundance were also observed at the transcriptional level. Regulation could be positive (*CAB*), or negative (pEA25, pEA207); it could be immediate and complete within one hour (CAB, pEA207) or begin only after a lag of four hours (pEA 25); and, finally, it could be LF only (pEA25), HF only (pEA207), or both LF and HF (*CAB*) (Marrs and Kaufman, 1989, 1991). Taken together, these studies indicate a blue light response system every bit as complex as that for phytochrome.

Studies with Arabidopsis

Feinbaum and Ausubel (1988) first demonstrated that the single *CHS* gene in *Arabidopsis* was inducible by white light, and a later study from the same laboratory (Feinbaum et al., 1991) demonstrated that the effect was a specific response to blue and UV-A light. Dark-grown transgenic seedlings containing a *CHS* promoter/*GUS* reporter gene construct were transferred to continuous blue, UV-A, or red light and GUS expression was analyzed. Whereas red light was only minimally inductive, both blue and UV-A light resulted in a strong increase in the level of the *GUS* mRNA (and a comparable increase in the mRNA of the endogenous *CHS* gene) when measured one or two days after the beginning of the light treatment. The blue light effect was apparently not a consequence of phytochrome phototransformation, as pulses of blue light given to seedlings grown under continuous red light caused a significant increase both in *GUS* and in endogenous *CHS* mRNA after 3 hours compared to the red light controls.

Kubasek et al. (1992) extended these studies to include other genes encoding enzymes in the phenylpropanoid pathway: phenylalanine ammonia lyase (*PAL1*), chalcone isomerase (*CHI*), and dihydroflavonol reductase (*DFR*). All of these mRNAs as well as that for *CHS* were strongly induced by blue light treatment of both dark-grown and red light-grown seedlings. The sequence of appearance of the various mRNAs following the onset of blue light treatment was first *PAL1*, next *CHS*, then *CHI*, followed by *DFR* – the same order as the encoded enzymes appear in the flavonoid biosynthesis pathway.

The above studies give no indication as to what blue light photoreceptor might be involved in these inductive responses. However, Chory (1992) reported that *hy4* seedlings accumulated less anthocyanin in white light than wild type, and two recent papers demonstrate an involvement of the CRY1 photoreceptor in the accumulation of anthocyanin. Some 16 separate *hy4* mutant alleles showed anthocyanin deficiencies during the first 6 days of growth under continuous blue light (Ahmad et al., 1995). *CHS* mRNA levels increased significantly in wild-type seedlings grown in red light and then transferred to blue light, whereas *hy4* seedlings grown under similar conditions showed no detectable change. Jackson and Jenkins (1995) showed that in addition to impaired induction of *CHS*, induction of *CHI*, and *DFR* transcripts by blue light was impaired as well. There is an extremely good correlation between the extent of blue light-dependent inhibition of hypocotyl growth and the impairment of induction of anthocyanin synthesis in response to blue light in 16 different *hy4* alleles (Ahmad et al., 1995). Together, studies provide convincing evidence for involvement of CRY1 in both photoresponses.

Jackson et al. (1995) have screened a mutagenized population of *Arabidopsis* seedlings expressing *GUS* under the control of the *CHS* promoter for seedlings exhibiting increased or decreased levels of blue light-induced *GUS* activity and several such mutant lines were isolated. Two of these exhibited elevated transcript levels not only for *GUS*, but also for the steady-state mRNA levels for *CHI, DFR*, and the endogenous *CHS* (*CAB* levels were unaffected). One of these mutants, named *icx1* (increased chalcone synthase expression) also exhibited pleiotropic effects on development, producing fewer leaf trichomes, and narrower leaves, and an altered seed coat. The authors conclude that ICX1 is involved in at least two processes occurring in the epidermis: namely light regulation of gene expression and at least some aspect of developmental patterning (Jackson et al., 1995).

As mentioned previously, the level of *CAB* mRNA is regulated both by phytochrome (Kaufman et al., 1984) and independently by blue light (Warpeha et al., 1989) in pea seedlings. Karlin-Neumann et al. (Karlin-Neumann et al., 1988) showed similar phytochrome photoregulation of mRNA levels for the *CAB1* (but not *CAB2* or *CAB3*) genes in *Arabidopsis*. Gao and Kaufman (1994) recently compared red and blue light regulation of *CAB1* mRNA levels in wild-type and *hy4* seedlings. They first established that *CAB1* could be regulated by blue light as well as red, and then showed that the blue light component of this regulation was independent of phyto-chrome phototransformation, with the light-induced increases showing different kinetics for red and blue light. Unlike the situation with the mRNAs for enzymes in the phenylpropanoid pathway, blue light regulation of the *CAB1* mRNA level in *hy4* was indistinguishable from that in wild-type. Thus CRY1 is not the photoreceptor mediating changes in *CAB1* mRNA abundance, but another independent blue light photoreceptor must be regulating the expression of this gene. The *icx1* mutation is without effect on *CAB* regulation by blue light, despite the obvious enhance-ment of the *CHS, CHI*, and *DFR* responses (Jackson et al., 1995).

The genes encoding enzymes of the phenylpropanoid pathway and nuclear-encoded genes for chloroplast proteins are not the only ones regulated by blue light in *Arabidopsis*. Kaldenhoff and co-workers isolated 14 cDNA clones from *Arabidopsis* seedlings that represented transcripts preferentially accumulating in blue light *versus* red light (Kaldenhoff et al., 1993). The single-copy gene corresponding to one of these clones, designated *AthH2*, encodes a protein of 286 amino acids (M_r of 30.5 kDa) with six membrane-spanning helices, characteristic of a water channel. Based on these characteristics, the authors transformed wild-type *Arabidopsis* with anti-sense *AthH2* and compared the osmotic properties of protoplasts from antisense plants with those from the wild-type progenitor (Kaldenhoff et al., 1995). Protoplasts from antisense plants exhi-bited dramatically slowed osmotic bursting, supporting the hypothesis that the protein encoded by AthH2 was a water channel.

The *AthH2* mRNA level was shown to increase rapidly and dramatically when seedlings grown under red light were transferred to blue light (Kaldenhoff et al., 1993). Abscisic acid also induced a strong increase in *AthH2* mRNA, but not heat shock or gibberellic acid. Studies using transgenic *Arabidopsis* expressing the *GUS* reporter gene under the control of the *AthH2* promoter indicated that its strongest expression was in elongating regions of roots and newly formed organs, as well as in stomata and neighboring cells (Kaldenhoff et al., 1995). *AthH2* message was detected specifically in guard cells of young plants by *in situ* analysis, while in more mature organs of *AthH2/GUS* transgenic plants, GUS activity was found predominantly near the vascular tissue. Cell fractionation studies and immunogold labeling both indicated that the AthH2 protein is localized to the plasma membrane.

Other plants

While blue and red light photoregulation of gene expression appear independent in pea and *Arabidopsis*, this is not always the case: In a phytochrome-deficient *aurea* mutant of tomato several genes photoregulated specifically by blue light in wild type show no induction whatsoever in *aurea* (Oelmülle et al., 1989). Hence in tomato, a response to blue light requires phytochrome phototransformation. Even in Arabdopsis, however, the signal transduction pathway for photoregulation of *CAB* by blue and red light must have common elements, despite having different photoreceptors. For example, etiolated *Arabidopsis det1* and *cop1* mutant seedlings develop as if they were growing in the light (Chory et al., 1989; Deng and Quail, 1992), and the proteins encoded by these loci apparently act as a negative regulators both of blue light photoreceptor-activated, and of phytochrome-regulated photomorphogenesis in the dark (Chory, 1992; Ang and Deng, 1994).

Fluhr and Chua (1986) showed ten years ago that photoregulation of the abundance of the mRNAs for two different ribulose-1,5-bisphosphate genes (*rcbS-3 A* and *rcbS-3C*) were under phytochrome control in young seedlings, but showed significant control by blue light in re-etiolated light-grown plants. This shift in regulation from phytochrome to a blue light photoreceptor was also found with the pea genes expressed transgenically in petunia. In a more recent paper describing the cloning and characterization of the *CHS* gene from mustard (*S. alba*), Batschauer et al. (1991) demonstrated that an increase in *CHS* mRNA abundance was strongly induced both by blue and by UV-A light in leaves of light-grown plants that were dark-adapted for three days. Subsequently, Frohnmeyer et al. (1992) examined the spectral sensitivity of *CHS* induction both in mustard and parsley (*Petroselinum crispum*). As in pea, regulation was largely if not entirely through phytochrome in young parsley seedlings, while older parsley seedlings showed a shift to a blue light photoreceptor and a loss of phytochrome regulation. Only blue and UV-A light were effective in re-etiolated light-grown parsley seedlings, and only UV-B in re-etiolated mustard seedlings. The absence of sensitivity to blue light in mustard in the Frohnmeyer et al. study is in contrast to that observed by Batschauer et al. and may be a reflection of the underlying developmental control of photoresponsiveness, given the very different dark adaptation times used (7 days in the Frohnmeyer study *versus* 3 days for Batschauer et al.).

In a detailed study on light-grown French bean plants (*Phaseolus vulgaris*), Sawbridge et al. (1994) presented convincing evidence that it is a blue light photoreceptor rather than phytochrome that regulates the mRNA levels for two different *rcbS* genes (*rbcS-1* and *rbcS-3*). They grew plants under low fluence-rate white light and then transferred them to equal fluence rates of red or blue light for two days prior to extracting RNA for northern analysis, or 12 h prior to extracting nuclei for run-on transcriptional analysis. Both genes were sharply up-regulated by blue light and virtually unaffected by red. Differential effects of the two light sources on photosynthesis were ruled out by the demonstration that photosynthesis was roughly the same at equal fluence rates from the red and blue light sources, and stomatal conductances were likewise the same. The authors calculated that the Pr/Pfr ratio was five times higher under the red light source than under the blue, but that phytochrome cycling rates were about equal. Again, it is clear that in light-grown plants, it is a blue light photoreceptor rather than phytochrome that controls transcript level for specific nuclear genes.

Perspective

Light regulation of gene expression can clearly be mediated by different photoreceptors. Phytochrome appears to predominate in the regulation of nuclear genes for chloroplast proteins early in development, but there is a developmental shift to blue light receptors as plants continue their development in the light. With genes such as *PAL, CHS, CHI*, and *DFR*, blue light regulation of expression in etiolated seedlings may be through CRY1 in *Arabidopsis*, but clearly some other photoreceptor regulates *CAB* expression. Though this section is supposedly dealing with signal transduction, the above studies really only deal with cause and ultimate effect. There is at present little information about the intervening steps, although the *det, cop*, and *icx* mutants may provide an entry to understanding some of the processes involved. The photophysiological studies on pea from Kaufman's laboratory tell us that there are HF and LF responses, that the kinetics of these responses are different for different genes, and that there are developmental differences between LF and HF responses as well. Are there different signal transduction pathways from a single photoreceptor? Are there multiple photoreceptors? Is there cross talk between systems? How do these systems interact with phytochrome? Blue light-induced growth inhibition in pea can be rapid and transient (Laskowski and Briggs, 1989) or slow and persistent (Warpeha and Kaufman,1989). Do these two types of responses occur at the level of gene expression? Do they share photoreceptors? It will ultimately be necessary to explore these sorts of photophysiological complexities in combination with the powers of genetics to address these questions.

Ion movements

While higher plants are immobile and hence unable to search out a favorable environment or escape from an unfavorable one, certain cells do have a system of electrogenic pumps and ion and water channels that allow these cells to change their turgor and hence their volume. These volume changes result in alterations in cell shape that are then used to drive appropriate responses to changes in environmental parameters such as temperature, humidity, and light. One such case is the swelling or shrinking of guard cells in response to environmental cues, causing opening or closing of stomata with important consequences for rates of CO_2 and O_2 exchange and water loss (Assmann, 1993). Another such case is the coordinated swelling and shrinking of large groups of opposed motor cells in the pulvini of a number of different plant species, leading to dramatic changes in leaf orientation, again of possible benefit to the plant in maximizing or minimizing the capture of radiant energy (Coté, 1995). Leaf movement is also a common mechanism to avoid excess heating in dry environments, and can serve to prevent photodamage when conditions are such that CO_2 uptake is limiting. In both stomata and pulvini, the uptake or lost of the major cation K^+ is the key to driving the water fluxes that ultimately lead to changes in cell shape.

It is not the purpose of this section to cover the detailed mechanisms by which stomatal aperture and pulvinar movement are regulated, but rather to examine what is known about how blue light regulates these movement responses. The reader is referred to several recent reviews for a more general consideration of stomatal movement (Assmann, 1993; Kearns and Assmann, 1993; see also Schroeder et al., 1994); a recent review by Coté (1995) on signal transduction in pulvinar movement, and a general review by Koller (1990) on leaf movements.

Guard cell responses

Although Zeiger and Hepler had demonstrated some 19 years ago that blue light could induce swelling of isolated guard cell protoplasts (Zeiger and Hepler, 1977) and it had been known for many years that a component of light-induced stomatal opening was blue-light specific, it was not until 1985 that the first careful quantitative studies on stomatal opening in response to blue light were done (Iino et al., 1985; Zeiger et al., 1985). Using an approach similar to that Kaufman and colleagues had applied to their studies of gene expression, Iino et al. (1985) placed plants under a sufficient fluence rate of red light to saturate photosynthesis both in the guard cells and in the underlying mesophyll as well as to saturate any other red light-mediated reactions that might influence stomatal aperture and then carried out careful time-course studies of the stomatal response to pulses of blue light by monitoring gas exchange. From the results of these experiments, they proposed that light converted an inactive component, A, into an active state, B, that led to stomatal opening. In this state, the stomata were refractory to further blue light excitation. However, in darkness, there was a thermal return of B to A, and concomitant recovery of light sensitivity, a course of events much like that occurring in phototropism (see Palmer et al., 1993b).

The patch-clamp studies by Assmann et al. (1985) directly demonstrated that blue light-induced a hyperpolarization of the plasma membrane of guard cell protoplasts. As the hyperpolarization measured in the protoplasts was more rapid than stomatal opening in response to blue light *in vivo*, it was a reasonable hypothesis that the electrophysiological change was related to the cellular swelling response. Hyperpolarization requires ATP, and hence is presumed to be electrogenic. The results presented by these workers (Assmann et al., 1985) supported a model in which blue light-induced extrusion of protons results in a hyperpolarization of the cell that consequently drives K^+ uptake, resulting in stomatal opening. However, the electron motive force- (emf) coupled K^+ uptake does not occur *via* a H^+/K^+ antiporter as the hyperpolarization can occur in the absence of external K^+.

Shimazaki et al. (1986) have shown that guard cell protoplasts held under continuous red light and given a pulse of blue light responded by extruding protons into the external medium. Because the process continues for many minutes after the end of the blue light pulse, it is unlikely that the light energy is directly driving a proton pump, in sharp contrast to the blue light-driven proton pump in the photosynthetic bacterium *Halobacterium halobium* (Stoeckenius and Bogomolni, 1982). The authors postulated blue light activation of an electrogenic plasma membrane H^+-ATPase as the cause of the proton extrusion.

Raghavendra questioned this model on two grounds (Raghavendra, 1990). First, initial efforts to inhibit blue light-induced proton extrusion with vanadate, a potent inhibitor of plasma membrane H^+-ATPases (Sze, 1984), were unsuccessful (Shimazaki et al., 1986). Second, blue light could be shown to stimulate a plasma membrane-localized redox system in guard cells (Vani and Raghavendra, 1989), a biochemical response fairly well characterized for a number of different plant tissues and likely involving a flavin as photoreceptor (see Asard and Caubergs, 1991). Hence, Raghavendra proposed a blue light-driven redox-based transmembrane flux of protons and electrons from inside to outside, as the source of acidification. Other studies (Gautier et al., 1992), showing that blue light against a red light background induced both acidification and O_2

uptake with a stoichiometry for $H^+:O_2$ near 4 for guard cell protoplasts, supported the redox model. The French group also showed that proton extrusion was greater in CO_2-free medium than medium at ambient CO_2 and suggested that the increase was a consequence of removal of competition for reducing equivalents between photosynthesis and the plasma membrane-localized redox system (Lascève et al., 1993). However, additional studies with vanadate have demonstrated strong inhibition both of stomatal opening in epidermal peels (Schwartz et al., 1991) and swelling of guard cell protoplasts (Amodeo et al., 1992). Apparently, the high concentrations of Cl^- present in the bathing medium had prevented vanadate uptake in the earlier study (Shimazaki et al., 1986). Hence, regardless of redox reactions, a plasma membrane-localized H^+-ATPase is likely involved in the blue light-mediated proton extrusion.

In efforts to elucidate other possible components of a blue light-activated signal transduction pathway in guard cells, Shimazaki and co-workers monitored proton extrusion by guard cell protoplasts in the presence of the various inhibitors reported to disrupt signal transduction in animal systems. Inhibitors of cyclic AMP- and cyclic GMP-dependent protein kinases, and of Ca^{2+}/calmodulin-dependent protein kinase II were without effect, and an inhibitor of protein kinase C had only a slight effect. However, inhibitors of Ca^{2+}-calmodulin-dependent myosin light chain kinase strongly suppressed blue light-dependent proton extrusion (Shimazaki et al., 1992). The authors suggested that the suppression observed was not directly on the H^+-ATPase itself, but rather on some earlier component in the signal transduction chain, as fusicoccin-stimulated proton efflux was unaffected. Given the caveats of experiments with inhibitors, these studies nevertheless present intriguing lines of investigation for future research.

At present, there is no definitive evidence as to the identity of the chromophore mediating blue light-induced stomatal opening. Recently, Srivastiva and Zeiger (1995) showed that dithiothreitol, an inhibitor of zeaxanthin formation (Yamamoto and Kamite, 1972), inhibits blue light-stimulated stomatal opening in epidermal peels of *Vicia faba*, and proposed that this carotenoid could be the photoreceptor. However, Karlsson et al. (1983) showed many years ago that the stomatal response to blue light was normal in wheat seedlings grown in the presence of the carotenoid synthesis inhibitor norflurazon. As a reducing agent such as dithiothreitol may have a wide range of non-specific effects, it seems likely that the inhibition of stomatal opening observed by Srivastiva and Zeiger was related to some process other than its effect on zeaxanthin formation.

It is unlikely that the photoreceptor for blue light effects on stomata is either CRY1 or NPH1. Chory (1992) and Liscum and Hangarter (1994) mention in passing that *Arabidopsis hy4* mutant alleles show normal stomatal regulation, and we have recently found that the *Arabidopsis nph1* mutants also show normal stomatal regulation (G. Lascève, A. Vavasseur, E. Liscum, W.R. Briggs, unpublished results).

Pulvinar movements

The literature on pulvinar movements is complicated by the fact that phytochromes, blue light photoreceptors, and underlying circadian effects all play roles in regulating turgor changes. Also, many potentially interesting studies on pulvinar movement have used white light only, making it impossible to assign any specific roles either to a blue-light photoreceptor or to phytochrome. In this section, we shall restrict ourselves to studies where it is clear that specific blue light-modulated processes are being studied. The involvement of phytochrome, circadian rhythms, and photosynthesis will be considered only in that context.

The current model for pulvinar movement (Coté, 1995) is not dissimilar to that for guard cells: turgor pressure is determined by an emf-coupled K^+ flux that is driven by a H^+-ATPase. The principle counterion entering with the K^+ is Cl^- although production of malate from starch may also play a role. Differential regulation of the H^+-ATPase in the flexor and extensor cells of the pulvinus results in differing turgors within these cells, thus resulting in movement. For example, when blue light strikes flexor cells from above, the activity of a H^+-ATPase is decreased while the activity of the H^+-ATPase in unchanged or perahps only slightly decreased in the extensor cells. Thus the flexor cells shrink with respect to the extensor cells, and the leaf is elevated.

Using intracellular electrodes in motor cells, Nishizaki (1986) first demonstrated a specific effect of blue light on ionic fluxes in pulvinar cells in *P. vulgaris*. Blue light directed at the flexor tissue induced a large and rapid change in membrane potential in the flexor cells, and a similar but smaller change in the extensor cells. Neither red nor green light had any measurable effect. The changes invariably preceded upward leaflet movement which started after a lag of 1–2 min. Nishizaki attributed movement to the differential fluxes between flexor and extensor, showing that changes in the membrane potential difference between flexor and extensor were closely related to rhythmic leaflet movements in studies extending over several hours. He subsequently showed that the blue light-induced change involved rapid membrane depolarization (Nishizaki, 1987).

More detailed photobiological studies with *P. vulgaris* (Nishizaki, 1988) demonstrated significant depolarization in response to brief (e.g., 15 s) pulses of blue light. Again, if the flexor was irradiated, the measured depolarization was greater in flexor than extensor cells, presumably a function of the steep light gradient across the pulvinus. A fluence-response curve showed saturation near 200 μmol m^{-2}, indicating that the system was very sensitive, in fact as sensitive as first positive phototropism (Iino, 1990).

Energy derived either from respiration or photosynthesis is clearly required to maintain the membrane potentials in motor cells of *P. vulgaris*. Anoxia in the dark collapses the potential and renders blue light ineffective, while red light during anoxia restores the potential and responsivity to blue light (Nishizaki, 1990). The inhibitor of photosynthetic electron transport DCMU (Nishizaki, 1992) prevents this red light-induced restoration of membrane potential. Both vanadate and dicyclohexylcarbodiimide (DCCD), inhibitors of H^+-ATPases, depolarized the motor cells, and blue light gave no further depolarization, supporting the participation of a proton pump.

Nishizaki (1988) adapted the Iino et al. model proposed for the guard cell response to blue light, in which light converts an inactive component to an active one and the active species then decays thermally in the dark back to the inactive form over a period of minutes (Iino et al., 1985) to explain the pulvinar response to blue light. In the case of pulvini, the presumed action of the active species is to *inhibit* the plasma membrane H^+-ATPase. Note that this effect is precisely the opposite of what is proposed to happen in guard cells where blue light is thought to *activate* a H^+-ATPase (Assmann et al., 1985).

Bialczyk and Lechowski (1990) examined the effects of 6 h exposures to red, blue, or far red light after a dark period on levels both of malate and of K^+ and Cl^- in flexor and extensor cells. Blue and far red light both essentially double malic acid levels in the extensor cells (away from the light) but have little effect on the flexor cells. Conversely, red light doubles the malic acid concentration in the flexor cells with little effect on the extensor cells. Both Cl^- and K^+ show similar patterns though the differences are not as large. Since phenylacetic acid, a known flavin antagonist (Hemmerich et al., 1967), inhibits the malic acid changes in blue light, they postulate a specific flavin-containing photoreceptor for the blue light effect. Phenylacetic acid had no effect on red-induced changes or on levels in darkness. Since the effects of red light in the flexor could be reversed by far red light (exposures of several hours), they propose that phytochrome operates in both flexor and extensor cells, but with opposite effects on malic acid concentration. The effect in extensor cells was postulated to be a typical phytochrome far red high-irradiance response (HIR) (see Kendrick and Kronenberg, 1994), and that in flexors a red HIR response, although the experiments do not rule out photosynthetic effects as well. Since the only significant blue light effects were seen in the extensor cells, they assume that the blue light photoreceptor is located only in these cells and not in the flexor cells. These results contradict those of Nishizaki, cited above, in which both tissues showed a response to blue light. However, as Bialczyk and Lechowski used light exposures of many hours, their experiments cannot be directly compared with the short-term experiments of Nishizaki.

Kim et al. (1992) isolated protoplasts from flexor and extensor regions of *Samanea saman* pulvini and measured light-induced changes in membrane potential using a dye-based assay. The authors found that white or blue light caused a transient hyperpolarization of the cells, whereas red light had no effect. Curiously, treatment with 200 mM K^+ following red light treatment caused rapid depolarization of flexor but not extensor cells, whereas the same treatment following white or blue light treatment caused rapid depolarization of extensor but not flexor cells. Since tetraethylammonium, a K$^+$-channel blocker, completely inhibited both depolarizations, K$^+$ channels were clearly implicated. The authors' conclusion was that these channels were closed in extensor cell and open in flexor cells in the dark. With white or blue light, channels in the flexor cells closed, and those in the extensor cells opened. As vanadate inhibited the blue light-induced hyperpolarization response, these experiments also implicated a H$^+$-ATPase.

These results directly contradict the observations made by Nishizaki, discussed above, in which blue light induced a clear differential depolarization of flexor and extensor cells in the pulvinus of *P. vulgaris*. They also contradict earlier results by Racusen and Satter (1975) in which white light induced depolarization in flexor tissue in *S. saman*. Kim et al. (1992) suggest that the cells *in planta* may behave differently than *in vitro*, but at this writing these contradictory results cannot be definitively reconciled.

The story is somewhat complicated by circadian rhythm studies with protoplasts indicating that in *S. saman*, flexor K$^+$ channels are open when extensor K$^+$ channels are closed and *vice versa*, with reciprocal changes occurring according to a circadian rhythm (Kim et al., 1993). If pulvinar protoplasts were transferred to darkness from the middle of the light period, flexor K$^+$ channels, normally closed, could be induced to open by brief pulses of red light. This effect could be completely reversed by far red light, clearly implicating phytochrome, and thus complicating the picture still further.

In addition to effects of K$^+$ fluxes (which are far from resolved), Ca^{2+} fluxes also appear to play a role in motor cell-driven leaf movements. Roblin et al. (1989) reported that the Ca^{2+} inhibitor lanthanum (La^{3+}) inhibited both red light-induced leaflet closure and blue light-induced leaflet opening in *Cassia fasciculata*, though considerably higher concentrations were required to inhibit the blue light effect. By contrast, two other Ca^{2+} channel blockers, verapamil and nifedipine, inhibited just the red light-induced closure. The authors interpret the results to mean that for red light-induced changes in Ca^{2+}, external stores are involved, whereas for blue light-induced changes, only internal stores are involved. Moyen et al. (1995) followed internal calcium concentrations in protoplasts isolated from *Mimosa pudica* pulvini irradiated with UV-A by measuring the fluorescence of a calcium-sensitive dye, INDO-1/AM, and found a transient doubling of internal Ca^{2+} within 3–4 min. The change was completely independent of external calcium stores, supporting the conclusion (Roblin et al., 1989) that blue light effects on cytoplasmic Ca^{2+} levels were mediated *via* internal Ca^{2+} stores. While the authors presumably were exciting a blue/UV-A-absorbing photoreceptor, the Pr form of phytochrome has an absorption band in the UV-A but not in the blue, and hence results with UV-A and blue are not necessarily directly comparable.

Morse et al. (1987) took a different approach, looking for changes in cellular components known to be involved in signal transduction cascades in animal cells. They demonstrated a small but significant rapid acceleration (within 15 s) of inositol phospholipid turnover in pulvini of *S. saman* in response to light. Subsequently, they showed changes in diacylglycerol in light, a small increase within 30 s followed by a small decrease after about 10 min (Morse et al., 1989). Unfortunately, the authors used white light throughout these studies. They briefly mention that the diacylglycerol response was also obtained in response to blue light, but do not show any data.

Although ionic movements are clearly involved in blue light-regulated pulvinar movements, there is still much that needs elucidation. Blue light photoreceptors and phytochrome may have different actions at different times in the circadian cycle; they may have different actions in intact flexor and extensor tissues; and responses measured in protoplasts may not necessarily reflect what is happening in the intact pulvini. Careful experimentation will be required to resolve the complexities of the pulvinar responses to light.

Stem elongation

Prior to their detailed studies on regulation of the abundance of various mRNAs by blue light, the Kaufman group had already demonstrated that blue light can induce both LF and HF responses in long-term growth extending over hours (Warpeha and Kaufman, 1989; Warpeha and Kaufman, 1990b). However, blue light also induces a rapid inhibition of stem growth, with a lag period of a minute or less (Meijer, 1968; Gaba and Black, 1979; Cosgrove, 1981). The growth rate decreases rapidly, and may attain levels as low as 30% of the initial growth rate in the dark within 30–40 min in continuous blue light (Laskowski and Briggs, 1989). Gaba and Black (1979) showed that blue light-induced and red light-induced inhibition of growth had significantly different kinetics and therefore were almost certainly mediated by different systems (the red light response was far slower), and Cosgrove (1985) used similar kinetic studies to demonstrate that the rapid inhibition of stem elongation and phototropism must be mediated by different systems (the onset of phototropism occurred much later than the onset of the rapid inhibition of growth), results that are now supported by genetic studies (Liscum et al., 1992; Liscum and Briggs, 1995). Finally, Laskowski and Briggs (1989) showed that the inhibition could be induced by pulses of blue light lasting only a few seconds in etiolated pea seedlings. Growth inhibition continued to increase for 10–15 min following the rapid inhibition of growth, but was not accompanied by any change in cell-sap osmotic pressure, turgor pressure, stem hydraulic conductance, or cell wall yield threshold, but rather by a decrease in the wall yielding coefficient (Cosgrove, 1988). Thus the effect of blue light is to reduce cell wall loosening, leading to the consequent inhibition of cell elongation. For a detailed discussion of these and other physiological and biophysical studies of light inhibition of growth, the reader is referred to the recent review by Cosgrove (1994).

Using both intracellular and extracellular recording techniques, Spalding and Cosgrove (1988) found that the rapid inhibition of growth by blue light in *Cucumis sativus* seedlings was preceded by a dramatic transient blue light-specific plasma membrane depolarization of the irradiated cells. The depolarization could be as much as 100 mV and recover to the dark value within 2–3 min. Furthermore, the depolarization was highly localized, and did not spread to adjacent tissues kept in the dark. Finally, at higher fluences rates, the lag periods both for inhibition of growth and depolarization decreased comparably.

Using isolated hypocotyl segments from *C. sativus*, Spalding and Cosgrove (1992) investigated the effects of various inhibitors on the depolarization response in *C. sativus*. Depolarization was not affected by Ca^{2+} channel blockers such as verapamil or La^{3+}, although these substances did reduce the rate of repolarization, suggesting a role for Ca^{2+} uptake during the recovery phase. The K^+ channel blocker tetraethylammonium was completely without effect. However, vanadate reduced or completely inhibited the blue light-induced depolarization. Hence, as was the case with pulvini, the action of blue light is likely to be the inactivation of a H^+-ATPase.

The depolarization effect of blue light can evidently be completely decoupled from the effect of blue light on elongation, as the elongation response is ordinarily weak or nonexistent in isolated sections (Cosgrove, 1981), although the depolarization response itself is just as great as in intact seedlings. By manipulating the medium, however, Shinkle and Jones (1988) were able to obtain reasonable blue light-induced inhibition of growth in sections. Under these conditions, the Ca^{2+} chelater EGTA abolished the response to blue light, again implicating Ca^{2+} in some way.

It is possible that at least a part of the depolarization response may be mediated through the CRY1 photoreceptor. In preliminary studies, Cho and Spalding (1995a, b) investigated this rapid electrical change both in wild-type *Arabidopsis* and in the *hy4* mutant. The magnitude of the response in the mutant was less than 30% that of wild type. Since detailed short-term growth kinetics for the *Arabidopsis* response to blue light have not been reported, it is not known in this case whether the change might be correlated to a rapid initial growth response, a long-term response such as that reported in pea (Warpeha and Kaufman, 1989), or both (or neither).

Leaf epidermis

Guard cells are not the only epidermal cells showing a hyperpolarization response to blue light. Staal et al. (1994) measured both proton secretion and membrane potential in epidermal peels of the *Argenteum* mutant of *P. sativum*. This mutant is advantageous in that the epidermis separates easily from the underlying mesophyll without major damage to the epidermal cells, allowing photobiological investigations in the absence of photosynthesis. These workers found that although both blue and red light induced outward proton pumping accompanied by hyperpolarization, the kinetics of the two responses were different, and the effects additive. The effect of red light was far red-reversible, while blue light effect was not. The photosynthesis inhibitor DCMU had no effect on the response to either wavelength region. Since the H^+-ATPase inhibitor DCCD completely inhibited proton pumping by either blue or red light, apparently both photoreceptor systems acted by activating the plasma membrane H^+-ATPase. The likely consequence of the proton pumping in this case is proposed to be cell wall acidification leading to cell wall loosening and consequent cell expansion. Note that whereas the electrophysiological response to blue light in the Staal et al. study (1994) is hyperpolarization of epidermal cells, that studied by Spalding and Cosgrove (1988) is depolarization, in parallel to the responses of guard cells and pulvinar motor cells discussed above.

Conclusion

The last decade has seen some exciting progress in our understanding of blue light signal transduction. The first higher plant blue light photoreceptor, CRY1, is now in hand. One can anticipate significant progress in elucidating the signal transduction elements participating downstream from photoexcited CRY1, particularly with respect to hypocotyl growth inhibition and the induction of the phenylpropanoid pathway. Prospects for further insights into the phototropism signal transduction pathway, and for identification of another blue light photoreceptor are also encouraging. Genetic studies tells us that plants undoubtedly have at least three different blue light photoreceptors and associated signal transduction systems, with little or no overlap of function between them. Clearly, mutants in stomatal responses to blue light and blue light-directed leaf movements would be of great help in making progress in these areas as well. The coming years should bring some exciting advances.

References

Ahmad, M. and Cashmore, A.R. (1993) *HY4* gene of *A. thaliana* encodes a protein with characteristics of a blue-light photoreceptor. *Nature* 366: 162–166.
Ahmad, M., Lin, C. and Cashmore, A.R. (1995) Mutations throughout an *Arabidopsis* blue-light photoreceptor impair blue-light-responsive anthocyanin accumulation and inhibition of hypocotyl elongation. *Plant J.* 8: 653–658.

Amodeo, G., Srivastava, A. and Zeiger, E. (1992) Vanadate inhibits blue light-stimulated swelling of *Vicia* guard cell protoplasts. *Plant Physiol.* 100: 1567–1570.

Ang, L.-H. and Deng, X.-W. (1994) Regulatory hierarchy of photomorphogenic loci: Allele-specific and light-dependent interaction between hy5 and cop1. *Plant Cell* 6: 613–628.

Asard, H. and Caubergs, R. (1991) LIAC activity in higher plants. *In*: F. Lenci, F. Ghetti, G. Columbetti, D.-P. Häder, and P.-S. Song (eds): *Biophysics of Photoreceptors and Photomovements in Microorganisms*. Plenum Publishing Co., New York, pp 181–189.

Assmann, S.A. (1993) Signal transduction in guard cells. *Ann. Rev. Cell Biol.* 9: 345–375.

Assmann, S.M., Simoncini, L. and Schröder, J.I. (1985) Blue light activates electrogenic ion pumping in guard cell protoplasts of *Vicia faba*. *Nature* 318: 285–287.

Bandurski, R.S. and Galston, A.W. (1951) Phototropic sensitivity of coleoptiles of albino corn. *Maize Genetic Coop. Newsletter*: 5.

Baskin, T.I. and Iino, M. (1987) An action spectrum in the blue and ultraviolet for phototropism in alfalfa. *Photochem. Photobiol.* 46: 127–136.

Batschauer, A. (1993) A plant gene for photolyase: an enzyme catalyzing the repair of UV-light-induced DNA damage. *Plant J.* 4: 705–709.

Batschauer, A., Ehmann, B. and Schäfer, E. (1991) Cloning and characterization of a chalcone synthase gene from mustard and its light-dependent expression. *Plant Mol. Biol.* 16: 175–185.

Batschauer, A., Gilmartin, P.M., Nagy, F. and Schäfer, E. (1994) The molecular biology of photoregulated genes. *In*: R.E. Kendrick and G.H.M. Kroneneberg (eds): *Photomorphogenesis in Plants*. Kluwer, Dordrecht, pp 559–599.

Berry-Lowe, S.L. and Meagher, R.B. (1985) Transcriptional regulation of a gene encoding the small subunit of ribulose-1, 5-bisphosphate carboxylase in soybean tissue is linked to the phytochrome response. *Mol. Cell Biol.* 5: 1910–1917.

Bialczyk, J. and Lechowski, Z. (1990) Effect of light quality on malic acid synthesis in *Phaseolus coccineus* pulvini. *Plant Physiology Biochem.* 28: 315–322.

Briggs, W.R. and Iino, M. (1983) Blue light-absorbing photoreceptors in plants. *Phil. Trans. R. Soc. Edinburgh* 303: 347–359.

Chamovitz, D.A. and Deng, X.-W. (1995) The novel components of the *Arabidopsis* light signaling pathway may define a group of general developmental regulators shared by both animal and plant kingdoms. *Cell* 82: 353–354.

Cho, M.H. and Spalding, E.P. (1995a). Blue light activates an anion channel to inhibit growth in *Arabidopsis* seedlings. *In*: *10th International Workshop on Plant Membrane Biology*. Regensburg, Germany.

Cho, M.H. and Spalding, E.P. (1995b). *Hy4* mediates the activation of anion channels by blue light in etiolated *Arabidopsis* hypocotyls. Abstract. *In*: *6th International Conference on Arabidopsis Research*. Madison, Wisconsin.

Chory, J. (1992) A genetic model for light-regulated seedling development in *Arabidopsis*. *Development* 115: 337–354.

Chory, J. (1993) Out of darkness: mutants reveal pathways controlling light-regulated development in plants. *Trends Genet.* 9: 167–172.

Chory, J. and Susek, R.E. (1994) *Light Signal Transduction and the Control of Seedling Development*. Cold Spring Harbor Press, Cold Spring Harbour, NY.

Chory, J., Peto, C.A., Ashbaugh, M., Saganich, R., Pratt, L. and Ausubel, F. (1989) Different roles for phytochrome in etiolated and green plants deduced from characterization of *Arabidopsis thaliana* mutants. *Plant Cell* 1: 867–880.

Cosgrove, D. (1981) Rapid suppression of growth by blue light: occurrence, time course, and general characteristics. *Plant Physiol.* 67: 584–590.

Cosgrove, D. (1985) Kinetic separation of phototropism from blue-light inhibition of stem elongation. *Photochem. Photobiol.* 42: 745–751.

Cosgrove, D.J. (1988) Mechanism of rapid suppression of cell expansion in cucumber hypocotyls after blue-light irradiation. *Planta* 176: 106–116.

Cosgrove, D.J. (1994) Photomodulation of growth. *In*: R.E. Kendrick and G.H.M. Kronenberg (eds): *Photomorphogenesis in Plants*. Kluwer, Dordrecht; pp 631–658.

Coté, G.G. (1995) Signal transduction in leaf movement. *Plant Physiol.* 109: 729–734.

Demmig-Adams, B. and Adams, W.W. (1992) Photoprotection and other responses of plants to high light stress. *Ann. Rev. Plant Physiol.* 43: 599–626.

Deng, X.-W. (1994) Fresh view of light signal transduction in plants. *Cell* 76: 423–426.

Deng, X.-W. and Quail, P.H. (1992) Genetic and phenotypic characterization of *cop1* mutants of *Arabidopsis thaliana*. *Plant J.* 2: 83–95.

Everett, M. and Thimann, K.V. (1968) Second positive phototropism in the *Avena* coleoptile. *Plant Physiol.* 43: 1786–1792.

Feinbaum, R.L. and Ausubel, F.M. (1988) Transcriptional regulation of the *Arabidopsis thaliana* chalcone synthase gene. *Mol. Cell. Biol.* 8: 1985–1992.

Feinbaum, R.L., Stortz, G. and Ausubel, F.M. (1991) High intensity and blue light regulated expression of chimeric chalcone synthase genes in transgenic *Arabidopsis thaliana* plants. *Mol. Gen. Genet.* 226: 449–456.

Fluhr, R. and Chua, N.-H. (1986) Developmental regulation of two genes encoding ribulose-bisphosphate carboxylase small subunit in pea and transgenic petunia plants: Phytochrome response and blue-light induction. *Proc. Natl. Acad. Sci. USA* 83: 2358–2362.

Frohnmeyer, H.B., Ehmann, B., Kretsch, T., Rocholl, M., Harter, K., Nagatani, A., Furuya, M., Bachauer, A., Hahlbrook, K. and Schäfer, E. (1992) Differential usage of photoreceptors for chalcone synthase gene expression during plant development. *Plant J.* 2: 899–906.

Furuya, M. (1993) Phytochromes: Their molecular species, gene families, and functions. *Ann. Rev. Plant Physiol.* 44: 617–645.

Gaba, V. and Black, M. (1979) Two separate photoreceptors control hypocotyl growth in green seedlings. *Nature* 278: 51–54.

Gallagher, S., Short, T.W., Ray, P.M., Pratt, L.H. and Briggs, W.R. (1988) Light-mediated changes in two proteins found associated with plasma membrane fractions from pea stem sections. *Proc. Natl. Acad. Sci. USA* 85: 8003–8007.

Galland, P. and Senger, H. (1988) The role of pterins in the photoreception and metabolism of plants. *Photochem. Photobiol.* 48: 811–820.

Galston, A.W. (1959) Phototropism of stems, roots and coleoptiles. *In*: W. Ruhland (ed.): *Handbuch der Pflanzenphysiologie*, Vol. XVII, part 1. Springer-Verlag, Berlin, pp 492–529.

Gao, J. and Kaufman, L.S. (1994) Blue-light regulation of the *Arabidopsis thaliana Cab1* gene. *Plant Physiol.* 104: 1251–1257.

Gautier, H., Vavasseur, A., G., L. and Boudet, A.M. (1992) Redox processes in the blue light response of guard cell protoplasts of *Commelina communis* L. *Plant Physiol.* 98: 34–38.

Gressel, J. (1979) Blue light photoreception. *Photochem. Photobiol.* 30: 749–754.

Hemmerich, P., Massey, V. and Weber, G. (1967) Photo-induced benzyl substitution of flavins by phenylacetate: a possible model for flavin catalysis. *Nature* 213: 728–730.

Iino, M. (1990) Phototropism: mechanisms and ecological implications. *Plant Cell Environ.* 13: 633–650.

Iino, M., Ogawa, T. and Zeiger, E. (1985) Kinetic properties of the blue-light response of stomata. *Proc. Natl. Acad. Sci. USA* 82: 8019–8023.

Jackson, J.A. and Jenkins, G.I. (1995) Extension-growth responses and expression of flavonoid biosynthesis genes in the *Arabidopsis hy4* mutant. *Planta* 197: 233–239.

Jackson, J.A., Fuglevand, G., Brown, B.A., Shaw, M.J. and Jenkins, G.I. (1995) Isolation of *Arabidopsis* mutants altered in the light-regulation of chalcone synthase gene expression using a transgenic screening approach. *Plant J.* 8: 369–380.

Jenkins, G.I., Christie, J.M., Fuglevard, G., Long, J.C. and Jackson, J.A. (1996) Plant responses to UV and blue light: biochemical and genetic approaches. *Plant Sci.* 112: 117–138.

Kaldenhoff, R., Kölling, A. and Richter, G. (1993) A novel blue light- and abscisic acid-inducible gene of *Arabidopsis thaliana* encoding an intrinsic membrane protein. *Plant Mol. Biol.* 23: 1187–1198.

Kaldenhoff, R., Kölling, A., Meyers, J., Karmann, U., Ruppel, G. and Richter, G. (1995) The blue light-responsive *AthH2* gene of *Arabidopsis thaliana* is primarily expressed in expanding as well as differentiating cells and encodes a putative channel protein of the plasmalemma. *Plant J.* 7: 87–95.

Karlin-Neumann, G.A., Sun, L. and Tobin, T.M. (1988) Expression of light-harvesting chlorophyll *a/b* protein genes is phytochrome-regulated in etiolated *Arabidopsis thaliana* seedlings. *Plant Physiol.* 88: 1323–1331.

Karlsson, E., Höglund, H.-O. and Klockare, R. (1983) Blue light induces stomatal transpiration in wheat seedlings with chlorophyll deficiency caused by SAN 9789. *Physiol. Plant.* 57: 417–421.

Kaufman, L.S. (1993) Transduction of blue-light signals. *Plant Physiol.* 102: 333–337.

Kaufman, L.S., Thompson, W.F. and Briggs, W.R. (1984) Different red light requirements for phytochrome-induced accumulation of *cab* RNA and *rbcS* RNA. *Science* 226: 1447–1449.

Kaufman, L.S., Watson, J.C., Briggs, W.R. and Thompson, W.F. (1985) Photoregulation of nucleus-encoded transcripts: blue-light regulation of specific transcript abundance. *In*: K.E. Steinback, S. Bonitz, C.J. Arntzen, and L. Bogorad (eds): *Molecular Biology of the Photosynthetic Apparatus*. Cold Spring Harbor Laboratory Press, Cold Spring Harbour, NY, pp 367–372.

Kearns, E.V. and Assmann, S.M. (1993) The guard cell-environment connection. *Plant Physiol.* 102: 711–715.

Kendrick, R.E. and Kronenberg, G.H.M. (1994) *Photomorphogenesis in Plants*. Kluwer Academic Publishers, Dordrecht, pp 828.

Kim, H.Y., Coté, G.G. and Crain, R.C. (1992) Effects of light on the membrane potential of protoplasts from *Samanea saman* pulvini. *Plant Physiol.* 99: 1532–1539.

Kim, H.Y., Coté, G.G. and Crain, R.C. (1993) Potassium channels in *Samanea saman* controlled by phytochrome and the biological clock. *Science* 260: 960–962.

Koller, D. (1990) Light-driven leaf movements. *Plant Cell Environ.* 13: 615–632.

Konjevic, R., Steinitz, B. and Poff, K.L. (1989) Dependence of the phototropic response of *Arabidopsis thaliana* on fluence rate and wavelength. *Proc. Natl. Acad. Sci. USA* 86: 9876–9880.

Konjevic, R., Khurana, J.P. and Poff, K.L. (1992) Analysis of multiple photoreceptor pigments for phototropism in a mutant of *Arabidopsis thaliana*. *Photochem. Photobiol.* 55: 789–792.

Koornneef, M. and Kendrick, R.E. (1994) Photomorphogenic mutants of higher plants. *In*: R.E. Kendrick and G.H.M. Kronenberg (eds): *Photomorphogenesis in Plants*. Kluwer Academic Publishers, Dordrecht, pp 601–628.

Koornneef, M., Rolff, E. and Spruit, C.J.P. (1980) Genetic control of light-inhibited hypocotyl elongation in *Arabidopsis thaliana* (L.) Heynh. *Zeitschrift für Pflanzenphysiologie* 100: 147–160.

Kubasek, W.L., Shirley, B.W., McKillop, A., Goodman, H.M., Briggs, W.R. and Ausubel, F.M. (1992) Regulation of flavonoid biosynthetic genes in germinating *Arabidopsis* seedlings. *Plant Cell* 4: 1229–1236.

Labouriau, L.G. and Galston, A.W. (1955) Phototropism in carotene-free plant organs. *Plant Physiol. (Suppl.)* 30: xxii.

Lascève, G., Gautier, H., Jappé, J. and Vavasseur, A. (1993) Modulation of the blue light response of stomata of *Commelina communis* by CO_2. *Physiol. Plant.* 88: 453–459.

Laskowski, M.J. and Briggs, W.R. (1989) Regulation of pea epicotyl elongation by blue light: Fluence response relationships and growth distribution. *Plant Physiol.* 89: 293–298.

Lin, C., Robertson, D.E., Ahmad, M., Raibekas, A.A., Dorns, M.S., Dutton, P.L. and Cashmore, A.R. (1995a). Association of flavin adenine dinucleotide with the *Arabidopsis* blue light receptor CRY1. *Science*: 968–970.

Lin, C., Ahmad, M., Gordon, D. and Cashmore, A.R. (1995b). Expression *Arabidopsis* crytochrome gene in transgenic tobacco results in hypersensitivity to blue, UV-A, and green light. *Proc. Natl. Acad. Sci. USA* 92: 8423–8427.

Liscum, E. and Briggs, W.R. (1995) Mutations in the *NPH1* locus of *Arabidopsis* disrupt the perception of phototropic stimuli. *Plant Cell* 7: 473–485.

Liscum, E. and Briggs, W.R. (1996) Mutations in transduction and response components of the phototropic signalling pathway. *Plant Physiol.*; in press.

Liscum, E. and Hangarter, R.P. (1991) *Arabidopsis* mutants lacking blue light-dependent inhibition of hypocotyl elongation. *Plant Cell* 3: 685–694.

Liscum, E. and Hangarter, R.P. (1994) Mutational analysis of blue-light sensing in *Arabidopsis*. *Plant Cell Environ.* 17: 639–648.

Liscum, E., Young, J.C., Poff, K.L. and Hangarter, R.P. (1992) Genetic separation of phototropism and blue light inhibition of stem elongation. *Plant Physiol.* 100: 267–271.

Lorenzi, R., Ceccarelli, N., Lercari, B. and Gualtieri, P. (1994) Identification of retinol in higher plants: Is a rhodopsin-like protein the blue light receptor? *Phytochem.* 36: 599–601.

Malhotra, K., Kim, S.T., Batschauer, A., Dawut, L. and Sancar, A. (1995) Putative blue-light photoreceptors from *Arabidopsis thaliana* and *Sinapis alba* with a high degree of sequence homology to DNA photolyase contain the two photolyase cofactors but lack DNA repair activity. *Biochem.* 34: 6892–6899.

Marrs, K.A. and Kaufman, L.S. (1989) Blue light regulation of transcription for nuclear genes in pea. *Proc. Natl. Acad. Sci. USA* 86: 4492–4495.

Marrs, K.A. and Kaufman, L.S. (1991) Rapid transcriptional regulation of the *Cab* and pEA207 gene families in peas by blue light in the absence of cytoplasmic protein synthesis. *Planta* 183: 327–333.

Meijer, G. (1968) Rapid growth inhibition of gherkin hypocotyls in blue light. *Acta Bot. Néerl.* 17: 9–14.

Millar, A.J., McGrath, R.B. and Chua, N.-H. (1994) Phytochrome Phototransduction Pathways. *Ann. Rev. Genet.* 28: 325–49.

Morse, M.J., Crain, R.C. and Satter, R.L. (1987) Light-stimulated inositol phospholipid turnover in *Samanea saman*. *Proc. Natl. Acad. Sci. USA* 84: 7075–7078.

Morse, M.J., Crain, R.C., Coté, G.G. and Satter, R.L. (1989) Light-stimulated inositol phospholipid turnover in *Samanea saman* pulvini. *Plant Physiol.* 89: 725–727.

Moyen, C., Cognard, C., Fleurat-Lessard, P., Raymond, G. and Roblin, G. (1995) Calcium mobilization under UV-A irradiation in protoplasts isolated from photosensitive pulvinar cells of *Mimosa pudica*. *J. Photochem. Photobiol.* 29: 59–64.

Nagatani, A., Reed, J.W. and Chory, J. (1993) Isolation and initial characterisation of *Arabidopsis* mutants that are deficient in phytochrome A. *Plant Physiol.* 102: 269–277.

Nishizaki, Y. (1986) Rhythmic and blue light-induced turgor movements and electrical potential in the laminar pulvinus of *Phaseolus vulgaris* L. *Plant Cell Physiol.* 27: 155–162.

Nishizaki, Y. (1987) Light-induced changes of electrical potential in pulvinar motor cells of *Phaseolus valgaris* L. *Plant Cell Physiol.* 28: 1163–1166.

Nishizaki, Y. (1988) Blue light pulse-induced transient changes of electric potential and turgor pressure in the motor cells of *Phaseolus vulgaris* L. *Plant Cell Physiol.* 29: 1041–1046.

Nishizaki, Y. (1990) Effects of anoxia and red light on changes induced by blue light in the membrane potential of pulvinar motor cells and leaf movement in *Phaseolus vulgaris* L. *Plant Cell Physiol.* 31: 591–596.

Nishizaki, Y. (1992) Effects of inhibitors on light-induced changes in electrical potential in pulvinar cells of *Phaseolus vulgaris* L. *Plant Cell Physiol.* 33: 1073–1078.

Oelmüller, R., Kendrick, R.E. and Briggs, W.R. (1989) Blue-light mediated accumulation of nuclear-encoded transcripts coding for proteins of the thylakoid membrane is absent in the phytochrome-deficient *aurea* mutant of tomato. *Plant Mol. Biol.* 13: 223–232.

Palmer, J.M., Short, T.W. and Briggs, W.R. (1993a). Correlation of blue light-induced phosphorylation to phototropism in *Zea mays* L. *Plant Physiol.* 102: 1219–1225.

Palmer, J.M., Short, T.W., Gallagher, S. and Briggs, W.R. (1993b). Blue light-induced phosphorylation of a plasma membrane-associated protein in *Zea mays* L. *Plant Physiol.* 102: 1211–1218.
Palmer, J.M., Warpheha, K.M.F. and Briggs, W.R. (1996) Zeaxanthin is not the photoreceptor for phototropism in maize coleoptiles. *Plant Physiol.* 110: 1323–1328.
Pang, Q. and Hays, J.B. (1991) UV-B-inducible and temperature-sensitive photoreactivation of cyclobutant pyrimidine dimers in *Arabidopsis thaliana. Plant Physiol.* 95: 536–543.
Parks, B.M. and Quail, P.H. (1991) Phytochrome-deficient *hy1* and *hy2* long hypocotyl mutants of Arabidopsis are defective in phytochrome chromophore biosynthesis. *Plant Cell* 3: 1177–1186.
Quail, P.H. (1991) Phytochrome: A light-activated molecular switch that regulates plant gene expression. *Ann. Rev. Genet.* 25: 389–409.
Quiñones, M.A. and Zeiger, E. (1994) A putative role of the xanthophyll, zeaxanthin, in blue light photoreception of corn coleoptiles. *Science* 264: 558–561.
Racusen, R. and Satter, R.L. (1975) Rhythmic and phytochrome-regulated changes in transmembrane potential in *Samanea* pulvini. *Nature* 255: 408–410.
Raghavendra, A.S. (1990) Blue light effects on stomata are mediated by the guard cell plasma membrane redox system distinct from the proton translocating ATPase. *Plant Cell Environ.* 13: 105–110.
Reymond, P., Short, T.W. and Briggs, W.R. (1992) Blue light activates a specific protein kinase in higher plants. *Plant Physiol.* 100: 655–661.
Roblin, G., Fleurat-Lessard, P. and Bonmort, J. (1989) Effects of compounds affecting calcium channels on phytochrome- and blue pigment-mediated pulvinar movements of *Cassia fasciculata. Plant Physiol.* 90: 697–701.
Rüdiger, W. and Briggs, W.R. (1995) Involvement of thiol groups in blue-light-induced phosphorylation of a plasma membrane-associated protein from coleoptile tips of *Zea mays*. L. *Zeit. Naturf.* 50c: 231–234.
Sancar, A. (1994) Structure and function of DNA photolyase. *Biochem.* 33: 2–9.
Sawbridge, T.I., López-Juez, E., Knight, M.R. and Jenkins, G.I. (1994) A blue-light photoreceptor mediates fluence-rate-dependent expression of genes encoding the small subunit of ribulose 1,5-bisphosphate carboxylase in light-grown *Phaseolus valgaris* primary leaves. *Planta* 192: 1–8.
Schroeder, J.I. (1994) Perspective on the physiology and structure of inward-rectifying K^+ channels in higher plants: Biophysical implications for K^+ uptake. *Ann. Rev. Biophys. Biomol. Struct.* 23: 441–471.
Schwartz, A., Illan, N. and Assmann, S.M. (1991) Vanadate inhibition of stomatal opening in epidermal peels of *Commelina communis. Planta* 183: 590–596.
Senger, H. (1980) *The Blue Light Syndrome.* Springer-Verlag, Berlin.
Senger, H. (1984) *Blue Light Effects in Biological Systems.* Springer-Verlag, Berlin.
Senger, H. (1987) *Blue Light Responses: Phenomena and Occurrence in Plants and Microorganisms,* Vol. I–II. CRC Press, Boca Raton, FL.
Shimazaki, K., Iino, M. and Zeiger, E. (1986) Blue light-dependent proton extrusion by guard-cell protoplasts of *Vicia faba. Nature* 319: 324–326.
Shimazaki, K., Kinoshita, T. and Nishimura, M. (1992) Involvement of calmodulin and calmodulin-dependent myosin light chain kinase in blue light-dependent H^+ pumping by guard cell protoplasts from *Vicia faba* L. *Plant Physiol.* 99: 1416–4121.
Shinkle, J.R. and Jones, R.L. (1988) Inhibition of stem elongation in *Cucumis* seedlings by blue light requires calcium. *Plant Physiol.* 86: 960–966.
Short, T.W. and Briggs, W.R. (1990) Characterization of a rapid, blue light-mediated change in detectable phosphorylation of a plasma membrane protein from etiolated pea (*Pisum sativum* L.) seedlings. *Plant Physiol.* 92: 179–185.
Short, T.W. and Briggs, W.R. (1994) The transduction of blue light signals in higher plants. *Ann. Rev. Plant Physiol.* 45: 143–171.
Short, T.W., Porst, M. and Briggs, W.R. (1992) A photoreceptor system regulating *in vivo* and *in vitro* phosphorylation of a pea plasma membrane protein. *Photochem. Photobiol.* 55: 773–781.
Short, T.W., Reymond, P. and Briggs, W.R. (1993) A pea plasma membrane protein exhibiting blue light-induced phosphorylation retains photosensitivity following triton solubilization. *Plant Physiol.* 101: 647–655.
Shropshire, W., Jr. and Mohr, H. (1983a). *Photomorphogenesis: Encyclopedia of Plant Physiology,* Vol. 16B. Springer-Verlag, Berlin.
Shropshire, W.J. and Mohr, H. (1983b). *Photomorphogenesis: Encyclopedia of Plant Physiology,* Vol. 16A. Springer-Verlag, Berlin.
Shropshire, W.J. and Withrow, R.B. (1958) Action spectrum of phototropic tip-curvature of *Avena. Plant Physiol.* 33: 360–365.
Silverthorne, J. and Tobin, E.M. (1984) Demonstration of transcriptional regulation of specific genes by phytochrome action. *Proc. Natl. Acad. Sci. USA* 81: 1112–1116.
Small, G.D., Min, B. and Lefebure, P.A. (1995) Characterization of a *Chlamydomonas reinhardtii* gene encoding a protein of the DNA photolyase/blue light photoreceptor family. *Plant Mol. Biol.* 28: 443–455.
Smith, H. (1995) Physiological and ecological function within the phytochrome family. *Ann. Rev. Plant Physiol.* 46: 289–315.

Spalding, E.P. and Cosgrove, D.J. (1988) Large plasma-membrane depolarization precedes rapid blue-light-induced growth inhibition in cucumber. *Planta* 178: 407–410.

Spalding, E.P. and Cosgrove, D.J. (1992) Mechanism of blue-light-induced plasma-membrane depolarization in etiolated cucumber hypocotyls. *Planta* 188: 199–205.

Srivastiva, A. and Zeiger, E. (1995) The inhibitor of zeaxanthin formation, dithiothreitol, inhibits blue light-stimulated stomatal opening in *Vicia faba*. *Planta* 196: 445–449.

Staal, M., Elzenga, J.T.M., van Elk, A.G., Prins, H.B.A. and Van Volkenburgh, E. (1994) Red and blue light-stimulated proton efflux by epidermal leaf cells of the Argenteum mutant of *Pisum sativum*. *J. Exp. Bot.* 45: 1213–1219.

Steinitz, B., Ren, Z. and Poff, K.L. (1985) Blue and green light-induced phototropism in *Arabidopsis thaliana* and *Lactuca sativa* L. seedlings. *Plant Physiol.* 77: 248–251.

Stoeckenius, W. and Bogomolni, R.A. (1982) Bacteriorhodopsin and related pigments in halobacteria. *Ann. Rev. Biochem.* 52: 587–616.

Sze, H. (1984) H^+-translocating ATPases of the plasma membrane and tonoplast of plant cells. *Physiol. Plant.* 61: 683–691.

Terzaghi, W.B. and Cashmore, A.R. (1995) Light-regulated transcription. *Ann. Rev. Plant Physiol.* 46: 445–474.

Thompson, W.F., Kaufman, L.S. and Watson, J.C. (1985) Induction of plant gene expression by light. *BioEssays* 3: 155–159.

Thompson, W.F. and White, M.J. (1991) Physiological and molecular studies of light-regulated nuclear genes in higher plants. *Ann. Rev. Plant Physiol.* 42: 423–466.

Tobin, E.M. (1978) Light regulation of specific mRNA species in Lemna gibba L. G-3. *Proc. Natl. Acad. Sci. USA* 75: 4749–4753.

Vani, T. and Raghavendra, A.S. (1989) Tetrazolium reduction by guard cells in abaxial epidermis of *Vicia faba*: blue light stimulation of a plasmalemma redox system. *Plant Physiol.* 90: 59–62.

Vierstra, R.D. and Poff, K.L. (1981) Role of carotenoids in the phototropic responses of corn seedlings. *Plant Physiol.* 68.

Warpeha, K.M.F. and Kaufman, L.S. (1989) Blue light regulation of epicotyl elongation in *Pisum sativum*. *Plant Physiol.* 89: 544–548.

Warpeha, K.M.F. and Kaufman, L.S. (1990a). Two distinct blue-light responses regulate the levels of transcripts of specific nuclear-coded genes in pea. *Planta* 182: 553–558.

Warpeha, K.M.F. and Kaufman, L.S. (1990b). Two distinct blue light responses regulate epicotyl elongation in pea. *Plant Physiol.* 92: 495–499.

Warpeha, K.M.F., Marrs, K.A. and Kaufmann, L.S. (1989) Blue-light regulation of specific transcript levels in *Pisum sativum*. *Plant Physiol.* 91: 1030–1035.

Whitelam, G.C. and Harberd, N.P. (1994) Action and function of phytochrome family members revealed through the study of mutant and transgenic plants. *Plant Cell Environ.* 17: 615–625.

Whitelam, G.C., Johnson, E., Peng, J., Carol, P., Anderson, M.L., Cowl, J.S. and Harberd, N.P. (1993) Phytochrome A null mutants of *Arabidopsis* display a wild-type phenotype in white light. *Plant Cell* 5: 757–768.

Yamamoto, Y. and Kamite, L. (1972) The effects of dithiothreitol on violxanthin deepoxidation and absorbance changes in the 500 nm region. *Biochim. Biophys. Acta* 267: 538–543.

Young, J.C., Liscum, E. and Hangarter, R.P. (1992) Spectral-dependence of light-inhibited hypocotyl elongation in photomorphogenic mutants of *Arabidopsis*: evidence for a UV photosensor. *Planta* 188: 106–114.

Zeiger, E. and Hepler, P.K. (1977) Light and stomatal function: blue light stimulates swelling of guard cell protoplasts. *Science* 196: 887–889.

Zeiger, E., Iino, M. and Ogawa, T. (1985) The blue light response of stomata: pulse kinetics and some mechanistic implications. *Photochem. Photobiol.* 42: 759–763.

Signal Transduction in Plants
P. Aducci (ed.)
© 1997 Birkhäuser Verlag Basel/Switzerland

The transduction of light signals by phytochrome

C. Bowler

Stazione Zoologica, Villa Comunale, I-80121 Napoli, Italy

Summary. Extraordinary progress has recently been made towards the understanding of how the plant photoreceptor phytochrome is able to transduce light signals into biological responses. This has been achieved by a combination of complementary genetic, biochemical, and cell biological approaches. We can now assign individual phytochromes to particular responses, and have knowledge about the nature and workings of some of the signal transduction intermediates utilized. This review summarizes our current knowledge of phytochrome phototransduction mechanisms.

Introduction

The absolute requirement for light during most aspects of growth is the key feature that distinguishes the Plant Kingdom from the Animal Kingdom. Light provides the energy source for photosynthesis and it also acts as an information source to tell the plant about its environment. Plant responses to light can be classified under photomorphogenesis, phototropism, and photoperiodism. Photomorphogenesis refers to the rapid changes that occur in a dark-grown "etiolated" seedling when it is exposed to light. The most apparent aspects of photomorphogenesis are stem shortening and thickening, apical hook opening, and cotyledon expansion accompanied by chloroplast biogenesis. Phototropism is defined as growth (e.g., stem bending) in response to directional light, whilst photoperiodism refers to responses (e.g., flowering) that are regulated by day length. Responses such as these allow the plant to synchronize developmental programmes with the acquisition of light for photosynthesis. For example, they can be utilized for proximity perception, for shade avoidance, and for the coordination of plant growth and development with the seasons.

Light is perceived by a range of photoreceptors, defined as being able to intercept light signals and to transduce the information to elicit appropriate responses. In higher plants, these photoreceptors comprise four classes, each of which is able to detect light of particular wavelengths: the phytochromes (responsive to red and far-red light), the blue light receptors, the ultraviolet-A (UV-A) and the ultraviolet-B (UV-B) receptors. Of these, the best characterized biochemically and physiologically is phytochrome.

Plants contain several phytochromes, e.g., five phytochrome genes (*PHYA – PHYE*) are present in *Arabidopsis* (Sharrock and Quail, 1989; Clack et al., 1994). Light absorption is mediated by a chromophore covalently bound to the apoprotein, and each phytochrome can exist in two photo-interconvertible forms, known as Pr and Pfr. The Pr form absorbs maximally red wavelengths (600 – 700 nm) whereas the Pfr form absorbs maximally in the far-red region (700 – 800 nm). Pfr is generally thought of as being the active form. A classical phytochrome-mediated response has therefore been defined as being promoted by a pulse of red light (which converts Pr to Pfr) and reversed by a pulse of far-red given subsequent to the red light flash (which converts Pfr back to Pr). Examples are seedling germination and the reduction of hypocotyl length in etiolated seedlings. In addition to such responses, known as low-fluence responses (LFR, requiring ~ 1 μmol/m^2 red light), phytochrome can elicit responses to continuous red or far-red irradiation, known as high irradiance responses (HIR, requiring > 10 μmol/m^2 light), and can mediate responses to extremely low-fluence red and far-red light, denoted very low-fluence responses (VLF, requiring <0.1 nmol/m^2 light) (Millar et al., 1994; Smith, 1995). The importance of a red/far-red sensor such as phytochrome for plant growth is that red:far-red ratios are excellent indicators of sun and shade environments (Smith, 1982). Phytochrome therefore plays a funda-mental role in the shade avoidance syndrome displayed by many plant species.

Phytochromes have been grouped principally into two classes, Type I and Type II. Type I phytochromes are light-labile whereas Type II phytochromes are light-stable. This difference is due to the fact that the Pfr form of Type I phytochrome is unstable compared to the Pr form, so that following irradiation with white or red light, Type I Pfr is rapidly degraded. Current evidence indicates that PHYA is the only Type I phytochrome, and that the remaining phytochromes are Type II (Smith, 1995).

Functions of individual phytochromes

Physiological functions for individual phytochromes have been largely deduced from studies of mutants that contain mutations in individual *PHY* genes. Most such mutants have been selected in *Arabidopsis* and their principal phenotype is that they are insensitive to light. Indeed, they were originally selected as mutants that grow in the light like a normal plant grows in the dark. Most conspicuously, they have long hypocotyls in the light (Koornneef et al., 1980), and consequently, they are known as *hy* mutants. Although the majority were originally described in 1980, only with the recent molecular identification of the genetic lesions has information been obtained about the physiological functions of individual phytochromes. A probable blue/UV-A photoreceptor has also been reported by identifying the molecular lesion in one of these mutants, *hy4* (Ahmad and Cashmore, 1993).

Mutants that lack both Type I and Type II phytochromes (e.g., *hy1*, *hy2*, and *hy6*) are deficient in chromophore biosynthesis. Although it may be expected that a mutation causing deficiency in all the phytochromes would be lethal, these mutants are viable, almost certainly because the mutations are leaky, inferred from the fact that some phytochrome-mediated phenomena can be observed in light grown plants (Whitelam and Smith, 1991). The phenotypes of the *phyB* mutants, originally denoted *hy3* (Quail et al., 1994), indicate that PHYB is principally responsible for the regulation of hypocotyl growth in constant red light and that it regulates classic red/far-red reversible responses. In contrast, the *phyA* mutant phenotypes (originally *hy8*, Quail et al., 1994) demonstrate that PHYA is necessary for continuous far-red light perception (known as the FR-HIR). Comparative analyses of *phyA*, *phyB*, and *phyAphyB* double mutants have suggested other roles for the individual phytochromes, but have also shown that some responses involve both, and that different members of the phytochrome family may be able to substitute for one another under certain situations (Reed et al., 1994; Smith, 1995).

The different red and far-red responses identified in the mutants are not due to different light-absorbing characteristics of PHYA and PHYB, but result from the fact that the Pfr form of PHYA (PfrA) is rapidly degraded in red light-containing light (half life 60–80 min). Consequently, in constant red light conditions, this rapid turnover of PHYA probably allows PHYB to assume the major role. Constant far-red light, however, ensures a steady photoconversion between PrA and PfrA, because Pr can be converted at low efficiency to Pfr in far-red light (2–3%) and can then be reconverted back to the Pr form rather than being degraded. Such a FR-HIR is particularly apparent in etiolated seedlings, because they contain about 50 times more PHYA than PHYB. Hence, PHYA responses such as the FR-HIR are probably important for the initial phases of seedling establishment, whereas the degradation of PfrA probably allows PHYB (and probably also the other Type II phytochromes) to play the critical roles in more mature green plants (see Quail et al., 1995, for review). No mutants have yet been reported that are specifically deficient in PHYC, D, or E.

The phytochrome molecule

Phytochrome is, to date, the only plant protein that has been unequivocally identified as a receptor. The majority of biochemical experiments have been carried out on monocot and pea PHYA due to the comparative ease with which sufficient amounts can be isolated from plant material compared with the Type II phytochromes. Sequence homologies between different Type I- and Type II-encoding genes suggest, however, that the basic biochemical properties of individual phytochromes will be very similar. A large body of evidence indicates that phytochromes are cytosolically-localized dimers consisting of two 125 kD polypeptides (~1150 amino acids) that each contain the same covalently-linked tetrapyrrole chromophore. Earlier data suggesting membrane or nuclear localizations (transient or otherwise) have not been further substantiated. Hence, although we know that it is a receptor, it does not look like any of the receptors so far characterized in animal cells. Furthermore, its ability to behave as a reversible on/off molecular light switch also sets it apart from the classical animal photoreceptor rhodopsin. These characteristics therefore provide no clues, *a priori*, as to how phytochrome is able to transduce light signals into appropriate biological responses, and this promise of discovering new molecular mechanisms makes phytochrome research one of the most stimulating and rewarding areas of modern plant research.

Due to the absence of any molecular models, structure-function studies have been carried out on modified phytochromes in an attempt to elucidate their mode of action. Unfortunately, many of the crucial questions about phytochrome activity have not been answered by such studies because experiments have largely been confined to the overexpression of mutant phytochrome gene sequences in wild-type plants, which of course contain the endogenous phytochromes. Nevertheless, the attachment site for the chromophore has been identified as an NH_2-terminal domain cysteine, and regions important for dimerization of the polypeptide have been identified in the centre and in the COOH-terminal domain (Edgerton and Jones, 1992; Quail et al., 1995).

Experiments in which NH_2- and COOH-terminal domains have been swapped in the *Arabidopsis PHYA* and *PHYB* genes suggest that the NH_2-terminal domain of PHYA is required to confer the FR-HIR, whereas the NH_2-terminal PHYB sequences confer sensitivity to constant red light (Quail et al., 1995). NH_2-terminal sequences therefore appear to be responsible for light detection, consistent with the known localization of the chromophore attachment site. By themselves, however, NH_2-terminal domains lack regulatory activity (Boylan et al., 1994). This, together with the fact that COOH-terminal domains of both PHYA and PHYB are able to function in conjunction with either NH_2-terminal domain suggest (1) that they carry the necessary information for the execution of phytochrome responses, and (2) that the mechanism of regulation is the same for both PHYA and PHYB (Boylan et al., 1994). Furthermore, removal of the COOH-terminal 35 residues from oat PHYA results in a protein that is physicochemically normal *in vitro* but which has no activity *in vivo* (Cherry et al., 1993). Hence, the COOH-terminal region is likely to play a key role in signal transduction. It has also been found, however, that the first NH_2-terminal 69 residues are important for signal transduction (Cherry et al., 1992;

Stockhaus et al., 1992). For example, multiple serine to alanine substitutions at the extreme NH_2-terminus of PHYA enhance its activity in transgenic plants (see later) (Stockhaus et al., 1992). To cloud the issue further, mutation of certain other NH_2-terminal residues (between amino acids 53 and 616) within the oat PHYA sequence result in a light-dependent dominant-negative phenotype when they are expressed in transgenic *Arabidopsis* (Boylan et al., 1994). Such a phenotype suggests that these mutated phytochrome molecules interact non-productively with the endogenous signaling components and that the mutated residues may be directly important for the interactions of PHYA with its downstream signaling components.

Sequence analysis of mutated phytochrome genes in *phyA* and *phyB* mutants of *Arabidopsis* has also yielded interesting structure-function information. The identification of such mutations is especially worthwhile because the nature of the mutation can be directly related to the severity of the mutant phenotype. Particularly interesting are mutations that inhibit phytochrome-mediated responses (such as hypocotyl growth inhibition) but that do not affect phytochrome photoreversibility. Interestingly, many such mutations in both *PHYA* and *PHYB* genes have been found to map to a 160-amino acid region from positions 680 to 840 and about 50% of such mutations (nine out of 21 examined) occur in just four residues between amino acids 776 and 793 (Quail et al., 1995). These residues are highly conserved in PHYA and PHYB sequences and are likely to be critical for mediating the contacts between the phytochrome molecules and their downstream signaling partners. The lack of homology of this region with any other sequences infers the involvement of a novel signaling mechanism which awaits future elucidation.

In summary, although several regions/residues can be directly implicated in the transduction of light signals from activated phytochrome, no one domain can be identified that is more important than the others. The fact that these sequences are scattered throughout the whole polypeptide highlights the necessity of obtaining a crystal structure of the phytochrome molecule with utmost urgency.

Data from animal cells have shown that signal transduction activity must be attenuated in order to terminate a response. Classical examples of attenuation (or desensitization) occur at the level of the receptor itself, e.g., rhodopsin is desensitized by phosphorylation (Kawamura, 1993). The observation that mutation of NH_2-terminal serine residues results in increased biological activity of PHYA (Stockhaus et al., 1992) suggests that phytochrome responses may also be desensitized by receptor phosphorylation, particularly because biochemical data have indicated that phytochrome is phosphorylated *in vivo* (Hunt and Pratt, 1980).

Downstream signal transduction components

Biochemical studies

Biochemical experiments aimed at elucidating the signaling cascades utilized by activated phytochromes have been based on the following approaches: (1) utilization of pharmacological agonists and antagonists to interfere with phytochrome-mediated events, (2) study of phosphorylation reactions or the effects of antibodies that inhibit signal transduction pathway components in animal cells, and (3) the use of microinjection to deliver signaling intermediates directly into the cells of a phytochrome-deficient tomato mutant.

The approaches listed here have identified G proteins as the most upstream component of the phytochrome signaling machinery. This was first inferred from studies of red light-induced swelling of wheat protoplasts (Bossen et al., 1990). It was shown that electroporation of protoplasts with the G protein activator GTPγS resulted in swelling in darkness, whereas electroporation of the G protein inhibitor GDPβS blocked the normal red light-induced swelling. In other experiments, treatment of dark-adapted photomixotrophic soybean cell cultures with cholera toxin resulted in the induction of *CAB* (clorophyll a,b-binding protein) gene expression (Romero and Lam, 1993), thus suggesting that G proteins of the heterotrimeric type were involved in mediating the normal light-dependent induction of this gene.

These results have been strengthened by experiments using a recently developed microinjection system that directly delivers signaling agonists and antagonists into hypocotyl cells of the phytochrome-deficient *aurea* mutant of tomato (Bowler and Chua, 1994; Neuhaus et al., 1993). These cells lack normal phytochrome-mediated responses such as chloroplast biogenesis, biosynthesis of photoprotectant anthocyanin pigments, and gene photoregulation. However, injection of purified oat PHYA can restore these responses (Neuhaus et al., 1993). Interestingly, cells injected with GTPγS or cholera toxin displayed identical responses to those stimulated with PHYA, whereas coinjection of PHYA with GDPβS or pertussis toxin, a heterotrimetic G protein inhibitor, resulted in the inhibition of the normal PHYA-mediated responses (Bowler et al., 1994a; Neuhaus et al., 1993). Furthermore, injection of cholera toxin or GTPγS in darkness stimulated the activation of light regulated genes such as *CAB* and *CHS* (which encodes the anthocyanin biosynthetic enzyme chalcone synthase), as assayed by coinjected *GUS* (β-glucuronidase) reporter gene constructs containing *CAB* and *CHS* promoters (Bowler et al., 1994a; Neuhaus et al., 1993). These results indicate that there are no light-requiring steps downstream of G protein activation for the induction of these genes. Hence, all light-requiring steps must be upstream of the G protein(s). Most likely then, the only light requiring step for the activation of these genes is that of phytochrome photoconversion.

By analogy with animal signal transduction schemes, we could therefore speculate that the initial event in phytochrome signal transduction may be the activation of a heterotrimeric G protein, a typical example of which is the activation of the G protein transducin by rhodopsin in animal photoreceptor cells (Koutalos and Yau, 1993; Stryer, 1991). There are considerable problems with this hypothesis, however, not least of which is the cellular localization of the molecules involved. Rhodopsin is an integral membrane protein and it is able to interact with transducin because it is also localized to the plasma membrane, as are all other known heterotrimeric G proteins. As previously mentioned, phytochrome is a cytosolic protein and there is no good evidence to support its transient localization to the membrane upon activation. Hence, either the visualization of such a phenomenon will only be revealed by more rigorous testing utilizing modern cell biological single cell imaging technology, or, alternatively, an intermediary molecule may transduce the signal from cytosolic Pfr to the membrane-localized G protein. Otherwise, it is possible that the G protein with which phytochrome interacts is not a heterotrimeric G protein in the classical "animal" sense, but is a novel cytoplasmically-localized type which maintains the pharmacological characteristics of the animal G proteins. Indirect support for the existence of a novel type of G protein in plants is that, even though several pharmacological studies have implicated G proteins in a range of responses (e.g., Fairley-Grenot and Assmann, 1991), to date only one gene has been cloned from plants (Ma et al., 1990). This contrasts strikingly with animal cells, which have at least 16 different genes (Neer, 1995).

Downstream of G protein activation, phytochrome utilizes calcium and cyclic GMP (cGMP) as second messengers (Bowler et al., 1994a; Neuhaus et al., 1993). Evidence for the involvement of calcium in phytochrome signal transduction has accumulated over the last ten years, largely through indirect experiments based on pharmacological manipulation of phytochrome-mediated fern spore germination, protoplast swelling, and gene expression (reviewed in Millar et al., 1994; Tretyn et al., 1991). Only comparatively recently, however, has a red/far-red-reversible Ca^{2+} transient been directly visualized in plant cells using calcium-sensitive fluorescent dyes (Shacklock et al., 1992). Microinjection of Ca^{2+} and Ca^{2+}-activated calmodulin (Ca^{2+}/CaM) into *aurea* cells subsequently demonstrated that these molecules are specifically involved in controlling PHYA-mediated chloroplast development (Neuhaus et al., 1993).

The same microinjection-based approach has revealed that cGMP controls PHYA-dependent anthocyanin biosynthesis and that it participates together with Ca^{2+} and Ca^{2+}/CaM to coordinate chloroplast development (Bowler et al., 1994a). These findings have been corroborated in an independent experimental system studying endogenous gene expression in photomixotrophic soybean cell cultures (Bowler et al., 1994a,b). The utilization of Ca^{2+} and cGMP to mediate phototransduction in plants has an interesting parallel with the photoperceptory mechanisms of animal photoreceptor cells, even though the photoreceptors utilized in each case (phytochrome and rhodopsin) are structurally completely unrelated (Koutalos and Yau, 1993).

Current biochemical knowledge of PHYA signal transduction pathways deduced from *aurea* microinjection and soybean cell culture pharmacology experiments can be summarized by the model in Figure 1a. Most interestingly, the targets of the three different signal transduction pathways are divided very clearly along functional lines. The cGMP pathway is utilized for the production of anthocyanin pigments, whereas the Ca^{2+}- and Ca^{2+}/cGMP-dependent pathways control different aspects of chloroplast development by regulating the expression of the genes . encoding subunits of the photosynthetic complexes. Coordination through these pathways appears to be controlled by two oppositely-acting negative regulatory pathways that have been termed reciprocal control (Bowler et al., 1994b) (Figs 1b and 1c). In one scenario, high activity of the Ca^{2+}-dependent pathways inhibits the cGMP-dependent pathway (Fig. 1b), whereas in the other, high activity of the cGMP-dependent pathway inhibits the Ca^{2+}-dependent pathways (Fig. 1c). Differential regulation of the two cGMP-dependent pathways is possible because they have different concentration requirements for cGMP: data from microinjection experiments indicate that the cGMP-dependent pathway controlling anthocyanin biosynthesis requires at least 30 μM cGMP, whereas the Ca^{2+}/cGMP-dependent pathway is operative even in the presence of 5 μM cGMP (as long as sufficient Ca^{2+} or Ca^{2+}/CaM are present, of course) (Bowler et al., 1994b). Hence, all the photosynthetic complexes can be synthesized even if the cGMP-dependent anthocyanin biosynthetic pathway is not operative.

We have proposed that such regulation is important at different times during plant development. Negative regulation of the Ca^{2+} and Ca^{2+}/cGMP pathways by high levels of cGMP (Fig. 1c) may be important during the early stages of light exposure in young seedlings to suppress chloroplast biogenesis at a time when there are not sufficient photoprotectants. This is consistent with the response known as juvenile anthocyanin (Drumm-Herrel, 1984). Negative regulation of the cGMP pathway by the Ca^{2+} pathways (Fig. 1b), on the other hand, suppresses anthocyanin

Figure 1. Summary of current biochemical models of PHYA signal transduction (adapted from Bowler et al., 1994b). (a) Design of the three basic signaling pathways dependent on cGMP and/or calcium. The cGMP-dependent pathway is responsible for activation of the genes required for anthocyanin biosynthesis, whereas the Ca^{2+}- and the Ca^{2+}/cGMP-dependent pathways activate genes required for the synthesis and assembly of the photosynthetic complexes. (b) Negative regulation of the cGMP pathway by the calcium pathways. (c) Negative regulation of the calcium- and calcium/cGMP-dependent pathways by the cGMP pathway. In (b) and (c), the dashed lines indicate the negative interactions, and brackets indicate the positions of the components which are responsible, based on current information (Bowler et al., 1994b). Abbreviations: Cyt. b_6f; Cytochrome b_6f, LHCI/II; light-harvesting complexes of PSI and PSII, PSI; Photosystem I, PSII; Photosystem II, RUBISCO; ribulose 1,5-bisphosphate carboxylase (see next page).

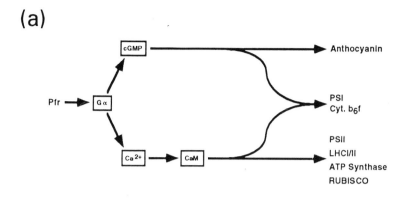

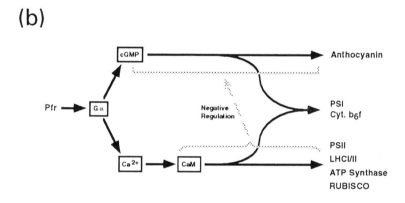

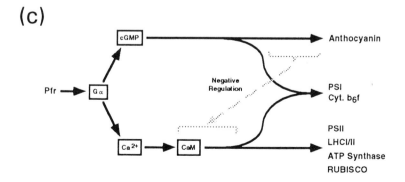

biosynthesis when they are no longer required and promotes chloroplast biogenesis. Such hypotheses are supported by the timing of expression of genes encoding anthocyanin biosynthetic enzymes and chloroplast components in soybean cells (Bowler et al., 1994b) and in seedlings exposed to red or white light (Brödenfeldt and Mohr, 1988; Drumm-Herrel, 1984; Ehmann et al., 1991; Pepper et al., 1994; Wenng et al., 1990).

The schemes discussed above demonstrate how complex outputs can be obtained from a relatively simple pre-programmed signaling circuitry. It is highly probable, therefore, that other responses can be integrated into the same model. For example, the fact that Photosystem II (PSII) and Photosystem I (PSI) complex biogenesis and assembly are controlled by different pathways (Fig. 1a) theoretically provides a means by which long-term adjustments in photosystem stoichiometry can be made in response to differing light conditions during growth, in order to retain high quantum efficiencies of photosynthesis in both sun and shade environments (Anderson, 1986; Melis, 1991).

Interestingly, diurnal oscillations in cytosolic calcium levels have recently been detected that are regulated by the circadian clock (Johnson et al., 1995). Some genes regulated by the Ca^{2+}-dependent PHYA signaling pathways are also under circadian control, e.g., *CAB* (Millar et al., 1992). Clearly then, a distinct possibility is that the circadian clock either generates or is regulated by these changing concentrations of cytosolic calcium and that this provides the molecular basis for the well known interaction between phytochrome and the clock. It will be interesting to examine calcium oscillations in phytochrome mutants and in the recently described *Arabidopsis* circadian clock mutants (Millar et al., 1995), as such experiments can potentially reveal whether calcium serves as an input or an output of the circadian oscillator. It would also be of great interest to determine whether cGMP levels exhibit circadian rhythms. The exquisite control of *CHS* induction and desensitization by changing levels of cGMP has already been documented (Bowler et al., 1994b); hence, diurnal changes may also exist that act together with the calcium oscillations to fine tune circadian clock activity and/or its subsequent outputs.

Genetic studies

Another approach for studying phytochrome signal transduction mechanisms has been to isolate mutants with defective photomorphogenesis. *Arabidopsis* is well suited to such studies because it exhibits typical photomorphogenic responses and has many features that distinguish it as a model plant system for molecular genetic studies. Two basic types of photomorphogenic mutants have been isolated (Chory, 1993): (1) those that have a partial etiolated morphology when grown in the light (e.g., a long hypocotyl), and (2) those that show characteristics of light-grown plants when grown in the dark (e.g., short hypocotyls, open cotyledons, and expression of light-regulated genes). In simple terms, the former class can be thought of as light-insensitive (i.e., blind) mutants, whereas the latter can be thought of as mutants constitutive for light responses.

Recessive light-insensitive mutants will most likely carry mutations in positively-acting components of the light signal transduction machinery. For example, the *hy* mutants previously described display classic light-insensitive phenotypes and are, indeed, blocked in positive signaling components, i.e., the photoreceptors themselves. Surprisingly, very few of the insensitive *hy* mutants so far isolated appear to be defective in the downsteam components of phytochrome signal transduction. *hy5* (Chory, 1992), *fhy1* and *fhy3* (Whitelam et al., 1993) are the only candidates reported to date. Such a paucity of mutants defective in positive elements downstream of phytochrome suggests that they are extremely difficult to find in these screens. If different phytochromes converge to utilize the same signaling molecules it is possible that mutation of these molecules would be lethal. Observations that PHYA and PHYB do have some overlapping functions (Reed et al., 1994) would indeed argue that at least some signaling components are shared, but nevertheless, the *fhy1* and *fhy3* mutants appear to specifically affect PHYA responses (Johnson et al., 1994; Whitelam et al., 1993). Whether the Ca^{2+} and cGMP pathways that have been defined biochemically (Fig. 1) are used specifically by PHYA or are shared by other phytochromes/photoreceptors remains to be determined.

Constitutive mutants have been isolated in screens for seedlings showing light-grown morphologies in darkness (discussed in Chory, 1993; Deng, 1994). Those isolated by Chory and co-workers are denoted *det* (for *det*iolated) whereas those described by Deng have been termed *cop* (for *co*nstitutive *p*hotomorphogenic). The recessive nature of these mutants can be most simply explained if they are loss of function alleles and that the wild-type gene products are negative regulators of photomorphogenesis. Epistasis tests using *hy* mutants have been used to propose that the gene products function downstream of the phyotchromes and the blue/UV-A photoreceptors (Chory, 1993; Deng, 1994) although this is difficult to conclude from such expeiments because of the individual phenotypes of the insensitive and constitutive mutants in dark and light, and becasue of the possible leakiness of the alleles being tested (see Millar et al., 1994, for a thorough discussion of these caveats). Furthermore, the light-grown phenotypes of the constitutive mutants indicate that they may be affected in other signaling pathways in addition to those controlling photomorphogenesis (Millar et al., 1994). This is a general difficulty of constitutive mutants – to be sure that the mutant is blocked in the signal transduction pathway of interest, it is necessary to show that *all* of the phenotypes of the mutant are consistent with *all* or at least *a subset of* the normally observed phenotype of wild-type plants exposed to the stimulus (discussed in Bowler and Chua, 1994). The fact that photomorphogenesis is regulated by so many inputs – different photoreceptors, as well as cytokinin and abscisic acid (Millar et al., 1994) – indicates that the assignment of a constitutive photomorphogenic mutant to a specific input signal will be extremely difficult to demonstrate. *det1*, for example, can be phenocopied in wild-type seedlings by addition of cytokinin (Chory et al., 1991; Chory et al., 1994).

Furthermore, six of the 11 constitutive mutants characterized to date (*det1, cop1, cop8, cop9, cop10*, and *cop11*) are allelic to the previously isolated fusca (*fus*) mutants, developmental mutants displaying seedling lethality (Castle and Meinke, 1994; McNellis et al., 1994a; Miséra et al., 1994; Wei et al., 1994b). Again, therefore, this evidence suggests that the *cop* and *det* mutants are more pleiotropic than was originally suspected.

In spite of the problems of linking the constitutive mutants specifically with the phytochromes, their phenotypes imply that the mutations lie within negatively-acting global regulators that are able to control plant development by integrating signals from multiple stimuli, including light. Furthermore, because they are allelic to seedling lethal mutants, it has been proposed that these molecules are critically important during early seedling developmet (Castle and Meinke, 1994).

Four of these negative regulators have now been cloned: *FUS6* (*COP11*), *COP9* (*FUS7*), *DET1* (*FUS2*), and *COP1* (*FUS1*). FUS6 (Castle and Meinke, 1994), *COP9* (Wei et al., 1994a), and *DET1* (Pepper et al., 1994) encode entirely novel proteins. Whereas FUS6 and COP9 are probably cytoplasmically-localized, DET1 is a nuclear protein. Attempts to determine the function of these proteins are being actively pursued. It has been found that COP9 exists as a large complex whose function or stability requires COP8 and COP11 (but not COP1, COP10, DET1, or HY5), because the complex cannot be detected in *cop8* and *cop11* mutants (Wei et al., 1994a). This is some of the first evidence that different *COP/DET/FUS* gene products interact with each other, and indicates that at least COP8, COP9, and COP11 may act together to control particular cellular events.

COP1, on the other hand, contains a combination of recognizable structural motifs: a zinc-finger domain at the NH_2-terminus, a putative coiled-coil region, and a domain at the COOH-terminus with multiple WD-40 repeats homologous to those found within the β-subunits of heterotrimeric G proteins and which are important for assembling macromolecular complexes (Deng et al., 1992; McNellis et al., 1994a). Furthermore, it bears significant homology to the $TAF_{II}80$ subunit of *Drosophila* TFIID, required for RNA polymerase II transcription initiation (Dynlacht et al., 1993). COP1 would therefore be predicted to be a nuclear protein, like DET1. This has indeed been found to be the case. Chimeric *COP1-GUS* fusions revealed that COP1 is localized to the nucleus in the dark but that exposure to light causes the relocation of the fusion protein to the cytoplasm (von Arnim and Deng, 1994). It has therefore been proposed that COP1 acts inside the nucleus to repress transcription in the dark, and that light reverts this activity by somehow expelling it from the nucleus. Based on its homology to $TAF_{II}80$, COP1 may impose transcriptional repression on target promoters in the dark by somehow interfering with activity of the TFIID complex. The physical expulsion of COP1 from the nucleus in the light would then de-repress TFIID activity. Such a mechanism could in theory explain why mutations in the *COP1* gene generate such pleiotropic phenotypes. Alternatively, it and/or DET1 could act by remodelling the nucleosomes in darkness to organize repressive regions of chromatin (Cooper et al., 1994; Imbalzano et al., 1994; Kwon et al., 1994). Yeast TUP1 is one such global repressor and, interestingly, it shares some homology with COP1 (Cooper et al., 1994; Deng et al., 1992). In such scenarios, the observed pleiotropy seen in *cop1* and *det1* mutants can be easily explained, even though one would normally tend to consider interactions at (or close to) the DNA level to be highly specific in their nature.

Arguing against the role of COP1 nucleocytoplasmic partitioning as a critical switch between dark and light developmental programmes is the observation that its expulsion from the nucleus is a relatively slow process, requiring at least 12 hours following irradiation of etiolated seedlings (von Arnim and Deng, 1994). The inhibition of hypocotyl elongation by light, on the other hand, occurs on a time scale of minutes (10–90 min depending on the plant species examined). COP1 may therefore function not as a critical molecular switch but rather for the maintenance of developmental programmes that were set in motion earlier by other light-regulated events (von Arnim and Deng, 1994). Such a hypothesis is supported by the fact that overexpression of COP1 results in only a partial suppression of light-mediated development (McNellis et al., 1994b).

Conclusion

The discussion above summarizes the extraordinary progress that has been made in recent years towards the elucidation of phytochrome signal transduction pathway components, utilizing very different but nonetheless complementary biochemical, cell biological, and genetic approaches. The currrent lack of overlap between the biochemical (Fig. 1) and the genetic (Ang and Deng, 1994; Chory, 1992) models of phytochrome signaling is now the major challenge to address in the future. It will also be necessary to relate the activities of the signaling intermediates to their downstream targets, the transcription factors that control phytochrome-regulated gene expression. Although several such factors have now been characterized (reviewed in Terzaghi and Cashmore, 1995) no links to individual signal transduction pathways or molecules have yet been reported.

It is important to realize that the differences between the biochemical and the genetic signaling models do not imply that one model is correct and the other not. The biochemical pathways have defined mainly the positive regulatory elements whereas the gene products identified in mutant studies are likely to be repressors. Furthermore, the biochemical pathways are highly specific cellular reactions (leading to chloroplast development and anthocyanin biosynthesis) whereas genetic studies take into account a whole variety of developmental inputs that manifest themselves not only at the single cell level but also as whole plant responses.

The phenotype of the *cop* and *det* mutants suggest that the mutations affect both the Ca^{2+} and cGMP pathways, because both chloroplast development and anthocyanin biosynthesis are constitutively de-repressed. Because we do not believe there is any activity of these pathways in the dark (i.e., no "dark current") (see Bowler and Chua, 1994; Bowler et al., 1994b), we therefore have to propose that the *COP* and *DET* gene products act downstream of the second messengers and that both Ca^{2+} and cGMP are normally involved in inactivating their repressor activity in the light. Current information about COP1 and DET1 structure imply that they can indeed be considered as downstream molecules because they are both nuclear proteins that most likely regulate transcription.

Another possibility is that the *cop* and *det* mutations affect only Ca^{2+}-dependent responses and that the anthocyanin biosynthesis observed in these mutants is a secondary stress response caused by the extremity of the mutant condition. If this is so, we would predict that inactivation of COP and DET repressor activity would be mediated by Ca^{2+} and perhaps negatively-regulated by cGMP, according to the reciprocal control model discussed previously.

In addition to locating the *COP* and *DET* gene products in the biochemical pathways, it will be highly worthwhile to isolate new mutants more likely to be affected specifically within the Ca^{2+} and cGMP pathways, or blocked in some aspect of reciprocal control. Sufficient information on the workings of the different pathways now exists to be able to predict the phenotypes of such mutants and to speculate that they could be isolated by designing mutant screens to detect aberrant phenotypes in dynamic responses to light signals (see Bowler and Chua, 1994). If progress continues at the rate of recent years, a thorough understanding of how plants perceive, respond, and adapt to changing light signals will soon be available at the molecular level.

References

Ahmad, M. and Cashmore, A.R. (1993) *HY4* gene of *A. thaliana* encodes a protein with characteristics of a blue-light photoreceptor. *Nature* 366: 162–166.

Anderson, J.M. (1986) Photoregulation of the composition, function, and structure of thylakoid membranes. *Annu. Rev. Plant Physiol.* 37: 93–136.

Ang, L.-H. and Deng, X.-W. (1994) Regulatory hierarchy of photomorphogenic loci: allele-specific and light-dependent interaction between the *HY5* and *COP1* loci. *Plant Cell* 6: 613–628.

Bossen, M.E., Kendrick, R.E. and Vredenberg, W.J. (1990) The involvement of a G-protein in phytochrome-regulated, Ca^{2+}-dependent swelling of etiolated wheat protoplasts. *Physiol. Plant.* 80: 55–62.

Bowler, C. and Chua, N.-H. (1994) Emerging themes of plant signal transduction. *Plant Cell* 6: 1529–1541.

Bowler, C., Neuhaus, G., Yamagata, H. and Chua, N.-H. (1994a) Cyclic GMP and calcium mediate phytochrome phototransduction. *Cell* 77: 73–81.

Bowler, C., Yamagata, H., Neuhaus, G. and Chua, N.-H. (1994b) Phytochrome signal transduction pathways are regulated by reciprocal control mechanisms. *Genes Dev.* 8: 2188–2202.

Boylan, M., Douglas, N. and Quail, P.H. (1994) Dominant negative suppression of *Arabidopsis* photoresponses by mutant phytochrome A sequences identifies spatially discrete regulatory domains in the photoreceptor. *Plant Cell* 6: 449–460.

Brödenfeldt, R. and Mohr, H. (1988) Time-courses for phytochrome-induced enzyme levels in phenylpropanoid metabolism (phenylalanine ammonia-lyase, naringenin-chalcone synthase) compared with time-courses for phytochrome-mediated end-product accumulation (anthocyanin, quercetin). *Planta* 176: 383–390.

Castle, L.A. and Meinke, D.W. (1994) A *FUSCA* gene of *Arabidopsis* encodes a novel protein essential for plant development. *Plant Cell* 6: 25–41.

Cherry, J.R., Hondred, D., Walker, J.M. and Vierstra, R.D. (1992) Phytochrome requires the 6-kDa N-terminal domain for full biological activity. *Proc. Natl. Acad. Sci. USA* 89: 5039–5043.

Cherry, J.R., Hondred, D., Walker, J.M., Keller, J.M., Hershey, H.P. and Vierstra, R.D. (1993) Carboxy-terminal deletion analysis of oat phytochrome A reveals the presence of separate domains required for structure and biological activity. *Plant Cell* 5: 565–575.

Chory, J. (1992) A genetic model for light-regulated seedling development in *Arabidopsis*. *Devel.* 115: 337–354.

Chory, J. (1993) Out of darkness: mutants reveal pathways controlling light-regulated development in plants. *Trends Genet.* 9: 167–172.

Chory, J., Aguilar, N. and Peto, C.A. (1991) The phenotype of *Arabidopsis thaliana det1* mutants suggests a role for cytokinins in greening. *Symp. Soc. Exp. Biol.* 45: 21–29.

Chory, J., Reinecke, D., Sim, S., Washburn, T. and Brenner, M. (1994) A role for cytokinins in de-etiolation in *Arabidopsis*. *Plant Physiol.* 104: 339–347.

Clack, T., Mathews, S. and Sharrock, R.A. (1994) The phytochrome apoprotein family in *Arabidopsis* is encoded by five genes: the sequences and expression of *PHYD* and *PHYE*. *Plant Molec. Biol.* 25: 413–427.

Cooper, J.P., Roth, S.Y. and Simpson, R.T. (1994) The global transcriptional regulators, SSN6 and TUP1, play distinct roles in the establishment of a repressive chromatin structure. *Genes Dev.* 8: 1400–1410.

Deng, X.-W. (1994) Fresh view of light signal transduction in plants. *Cell* 76: 423–426.

Deng, X.-W., Matsui, M., Wei, N., Wagner, D., Chu, A.M., Feldmann, K.A. and Quail, P.H. (1992) *COP1*, an *Arabidopsis* regulatory gene, encodes a protein with both a zinc-binding motif and a Gβ homologous domain. *Cell* 71: 791–801.

Drumm-Herrel, H. (1984) Blue/UV light effects on anthocyanin synthesis. In: H. Senger (ed.): *Blue Light Effects in Biological Systems*, Springer-Verlag, Berlin, Heidelberg, pp 375–383.

Dynlacht, B.D., Weinzierl, R.O.J., Admon, A. and Tjian, R. (1993) The dTAF$_{II}$80 subunit of *Drosophila* TFIID contains β-transducin repeats. *Nature* 363: 176–179.

Edgerton, M.D. and Jones, A.M. (1992) Localization of protein-protein interactions between subunits of phytochrome. *Plant Cell* 4: 161–171.

Ehmann, B., Ocker, B. and Schäfer, E. (1991) Development- and light-dependent regulation of the expression of two different chalcone synthase transcripts in mustard cotyledons. *Planta* 183: 416–422.

Fairley-Grenot, K. and Assmann, S.M. (1991) Evidence for G-protein regulation of inward K$^+$ channel current in guard cells of fava bean. *Plant Cell* 3: 1037–1044.

Hunt, R.E. and Pratt, L.H. (1980) Partial characterization of undegraded oat phytochrome. *Biochem.* 19: 390–394.

Imbalzano, A.N., Kwon, H., Green, M.R. and Kingston, R.E. (1994) Facilitated binding of TATA-binding protein to nucleosomal DNA. *Nature* 370: 481–485.

Johnson, C.H., Knight, M.R., Kondo, T., Masson, P., Sedbrook, J., Haley, A. and Trewavas, A. (1995) Circadian oscillations of cytosolic and chloroplastic free calcium in plants. *Science* 269: 1863–1865.

Johnson, E., Bradley, M., Harberd, N.P. and Whitelam, G.C. (1994) Photoresponses of light-grown *phyA* mutants of *Arabidopsis*. *Plant Physiol.* 105: 141–149.

Kawamura, S. (1993) Rhodopsin phosphorylation as a mechanism of cyclic GMP phosphodiesterase regulation by S-modulin. *Nature* 362: 855–857.

Koornneef, M., Rolff, E. and Spruit, C.J.P. (1980) Genetic control of light-inhibited hypocotyl elongation in *Arabidopsis thaliana* (L.) Heynh. *Z. Pflanzenphysiol.* 100: 147–160.

Koutalos, Y. and Yau, K.-W. (1993) A rich complexity emerges in phototransduction. *Curr. Opin. Neurobiol.* 3: 513–519.

Kwon, H., Imbalzano, A.N., Khavari, P.A., Kingston, R.E. and Green, M.R. (1994) Nucleosome disruption and enhancement of activator binding by a human SWI/SNF complex. *Nature* 370: 477–481.

Ma, H., Yanofsky, M.F. and Meyerowitz, E.M. (1990) Molecular cloning and characterization of GPA1, a G protein α subunit gene from *Arabidopsis thaliana*. *Proc. Natl. Acad. Sci. USA* 87: 3821–3825.

McNellis, T.W., von Arnim, A.G., Araki, T., Komeda, Y., Miséra, S. and Deng, X.-W. (1994a) Genetic and molecular analysis of an allelic series of *cop1* mutants suggests functional roles for the multiple protein domains. *Plant Cell* 6: 487–500.

McNellis, T.W., von Arnim, A.G. and Deng, X.-W. (1994b) Overexpression of *Arabidopsis* COP1 results in partial suppression of light-mediated development: evidence for a light-inactivable repressor of photomorphogenesis. *Plant Cell* 6: 1391–1400.

Melis, A. (1991) Dynamics of photosynthetic membrane composition and function. *Biochem. Biophys. Acta* 1058: 87–106.

Millar, A.J., Short, S.R., Chua, N.-H. and Kay, S.A. (1992) A novel circadian phenotype based on firefly luciferase expression in transgenic plants. *Plant Cell* 4: 1075–1087.

Millar, A.J., McGrath, R.B. and Chua, N.-H. (1994) Phytochrome phototransduction pathways. *Annu. Rev. Genet.* 28: 325–349.

Millar, A.J., Carré, I.A., Strayer, C.A., Chua, N.-H. and Kay, S.A. (1995) Circadian clock mutants in *Arabidopsis* identified by luciferase imaging. *Science* 267: 1161–1163.

Miséra, S., Müller, A.J., Weiland-Heidecker, U. and Jürgens, G. (1994) The *FUSCA* genes of *Arabidopsis*: negative regulators of light responses. *Molec. Gen. Genet.* 244: 242–252.

Neer, E.J. (1995) Heterotrimeric G proteins: organizers of transmembrane signals. *Cell* 80: 249–257.

Neuhaus, G., Bowler, C., Kern, R. and Chua, N.-H. (1993) Calcium/calmodulin-dependent and -independent phytochrome signal transduction pathways. *Cell* 73: 937–952.

Pepper, A., Delaney, T., Washburn, T., Poole, D. and Chory, J. (1994) *DET1*, a negative regulator of light-mediated development and gene expression in *Arabidopsis*, encodes a novel nuclear-localized protein. *Cell* 78: 109–116.

Quail, P.H., Boylan, M.T., Parks, B.M., Short, T.W., Xu, Y. and Wagner, D. (1995) Phytochromes: photosensory perception and signal transduction. *Science* 268: 675–680.

Quail, P.H., Briggs, W.R., Chory, J., Hangarter, R.P., Harberd, N.P., Kendrick, R.E., Koornneef, M., Parks, B., Sharrock, R.A., Schäfer, E., Thompson, W.F. and Whitelam, G.C. (1994) Spotlight on phytochrome nomenclature. *Plant Cell* 6: 468–471.

Reed, J.W., Nagatani, A., Elich, T.D., Fagan, M. and Chory, J. (1994) Phytochrome A and phytochrome B have overlapping but distinct functions in *Arabidopsis* development. *Plant Physiol.* 104: 1139–1149.

Romero, L.C. and Lam, E. (1993) Guanine nucleotide binding protein involvement in early steps of phytochrome-regulated gene expression. *Proc. Natl. Acad. Sci. USA* 90: 1465–1469.

Shacklock, P.S., Read, N.D. and Trewavas, A.J. (1992) Cytosolic free calcium mediates red light-induced photomorphogenesis. *Nature* 358: 753–755.

Sharrock, R.A. and Quail, P.H. (1989) Novel phytochrome sequences in *Arabidopsis thaliana*: structure, evolution, and differential expression of a plant regulatory photoreceptor family. *Genes Dev.* 3: 1745–1757.

Smith, H. (1982) Light quality, photoperception, and plant strategy. *Annu. Rev. Plant Physiol. & Plant Mol. Biol.* 33: 481–518.

Smith, H. (1995) Physiological and ecological function within the phytochrome family. *Annu. Rev. Plant Physiol. Plant Mol. Biol.* 46: 289–315.

Stockhaus, J., Nagatani, A., Halfter, U., Kay, S., Furuya, M. and Chua, N.-H. (1992) Serine-to-alanine substitutions at the amino-terminal region of phytochrome A result in an increase in biological activity. *Genes Dev.* 6: 2364–2372.

Stryer, L. (1991) Visual excitation and recovery. *J. Biol. Chem.* 266: 10711–10714.

Terzaghi, W.B. and Cashmore, A.R. (1995) Light-regulated transcription. *Annu. Rev. Plant Physiol.* 46: 445–474.

Tretyn, A., Kendrick, R.E. and Wagner, G. (1991) The role(s) of calcium ions in phytochrome action. *Photochem. Photobiol.* 54: 1135–1155.

von Arnim, A.G. and Deng, X.-W. (1994) Light inactivation of *Arabidopsis* photomorphogenic repressor COP1 involves a cell-specific regulation of its nucleocytoplasmic partitioning. *Cell* 79: 1035–1045.

Wei, N., Chamovitz, D.A. and Deng, X.-W. (1994a) *Arabidopsis* COP9 is a component of a novel signaling complex mediating light control of development. *Cell* 78: 117–124.

Wei, N., Kwok, S.F., von Arnim, A.G., Lee, A., McNellis, T.W., Piekos, B. and Deng, X.-W. (1994b) *Arabidopsis* *COP8*, *COP10*, and *COP11* genes are involved in repression of photomorphogenic development in darkness. *Plant Cell* 6: 629–643.

Wenng, A., Batschauer, A., Ehmann, B. and Schäfer, E. (1990) Temporal pattern of gene expression in cotyledons of mustard (*Sinapis alba* L.) seedlings. *Bot. Acta* 103: 240–243.

Whitelam, G.C. and Smith, H. (1991) Retention of phytochrome-mediated shade avoidance responses in phytochrome-deficient mutants of *Arabidopsis*, cucumber and tomato. *J. Plant Physiol.* 139: 119–125.

Whitelam, G.C., Johnson, E., Peng, J., Carol, P., Anderson, M.L., Cowl, J.S. and Harberd, N.P. (1993) Phytochrome A null mutants of *Arabidopsis* display a wild-type phenotype in white light. *Plant Cell* 5: 757–768.

Signal Transduction in Plants
P. Aducci (ed.)
© 1997 Birkhäuser Verlag Basel/Switzerland

Perception of fungal elicitors and signal transduction

F. Cervone[1], R. Castoria[2], F. Leckie[1] and G. De Lorenzo[1]

[1] Dipartimento di Biologia Vegetale, Università di Roma La Sapienza, I-00185 Rome, Italy
[2] Dipartimento di Scienze Animali, Vegetali e dell'Ambiente, Università degli Studi del Molise,
I-86100 Campobasso, Italy

Introduction

Plants, like animals, are continually exposed to a vast array of potential fungal pathogens; in many cases, they resist attack by blocking fungal development soon after penetration. As plants lack a circulatory system and antibodies, they have evolved defense mechanisms that are distinct from the vertebrate immune system. In contrast to animal cells, each plant cell is capable of defending itself by means of a combination of constitutive mechanisms and induced responses. After the perception of the pathogen (recognition), the plant cell at the site of infection transmits the information inside the cell across the plasma membrane as well as to neighboring cells. As a consequence, a number of defense reactions are induced, which include a rapid and localized cell death (hypersensitive response), a rapid oxidative burst, cross-linking and strengthening of the plant cell wall, the induction of the phenylpropanoid pathway and synthesis of lignin, the accumulation of antimicrobial compounds named phytoalexins, the synthesis of hydroxyproline-rich glycoproteins (HRGPs) and fungal wall degrading enzymes (chitinases, glucanases), and the production of ethylene. The effectiveness of the plant defense responses against a pathogen depends both on the magnitude and on the rapidity of their onset (Dixon and Lamb, 1990).

Two kinds of resistance control infections by fungal pathogens: resistance at the species level (non-host specificity) and resistance at the cultivar level (race-cultivar specificity). It is widely believed that in both cases the recognition is mediated by signal molecules (elicitors) produced by the pathogen and complementary plant "receptor" molecules. Signals and receptors are thought to interact at the contact surfaces between the two organisms to start a signal transduction pathway leading to the activation of the various defense responses. In spite of intense research, there are only a few cases where the interactions can be defined in molecular terms.

Many different molecules have been shown to behave as elicitors of defense responses when applied to plant tissues (Hahn et al., 1989; Darvill et al., 1992; Ebel and Cosio, 1994; Boller,

1995). Virtually all surface or secreted fungal molecules involved in the early stages of the plant-pathogen interaction may potentially play a role in the recognition process and in the trigger of plant defense responses. Non-specific elicitors are those molecules that elicit defense responses in a large variety of plant species and, within the same plant species, in a large variety of cultivars. Cultivar-specific elicitors are instead those molecules, direct or indirect products of the avirulence (*Avr*) genes, which are active only on the cultivars having the corresponding resistance (*R*) genes.

The final stages of the signal transduction pathway leading to the transcriptional activation of several defense genes have been elucidated in some cases. Both *cis*-elements and *trans*-factors for defense gene activation have been identified (Dixon et al., 1994 and references therein). Recently, the role of various components known as second messengers in animal systems has been tentatively assessed.

This chapter will deal with some of the elicitors so far characterized, their binding proteins, and the possible transduction steps which from the elicitor signal lead to the activation of defense responses.

Elicitors: Perception and transduction

Pectic enzymes and oligogalacturonides

The polysaccharide-rich cell wall is one of the first lines of defense against phytopathogenic fungi. The vast majority of fungi need to breach this barrier to gain access to the plant tissue and to this purpose secrete a number of enzymes capable of degrading the polymers of the plant cell wall. Pectic enzymes with an "endo" mode of action have often been associated with elicitor activity. In particular, fungal *endo*polygalacturonases have been reported to elicit a number of defense responses as different as the accumulation of phytoalexins (Lee and West, 1981; Walker-Simmons et al., 1984; Favaron et al., 1992), proteinase inhibitor I (Walker-Simmons et al., 1984), lignin (Robertsen, 1987), peroxidases (Bruce and West, 1989), and β-1,3-glucanases (Lafitte et al., 1993). Elicitor activity of *endo*polygalacturonases suggests that during an attempted invasion the enzymes may play two antithetical roles: as efficient fungal aggression tools, or potential signal molecules (elicitors). The early timing of *endo*polygalacturonase production is compatible with both functions. The available evidence indicates that fungal *endo*polygalacturonases are not directly responsible for the induction of plant defense responses, but are rather "pre-elicitors" that release from the plant cell wall the true elicitors, the oligogalacturonides (Anderson, 1989).

The first observation that a galacturonide-rich fraction released from the walls induces accumulation of phytoalexins in soybean was reported by Hahn et al. (1981), who named the active component of the fraction "endogenous elicitor". Since then, oligogalacturonides have been shown to elicit a variety of defense responses, including the accumulation of phytoalexins (Davis et al., 1986), the synthesis of endo-β-1,3-glucanase (Davis and Hahlbrock, 1987) and chitinase (Broekaert and Peumans, 1988), the synthesis of lignin (Robertsen, 1986; Bruce and West, 1989) and elicitation of necrosis (Marinelli et al., 1991). In carrot suspension-cultured cells, oligogalacturonides induce the activation of PAL, a key enzyme in phenylpropanoid and lignin biosynthesis (De Lorenzo et al., 1987). Responses elicited in carrot cells by oligogalacturonides appear to be a consequence of a transcriptional activation of several defense genes (Messiaen and Van Cutsem, 1993).

Only oligalacturonides of chain length varying between 10 and 15 are elicitors of defense responses; shorter oligomers usually have little or no elicitor activity (Hahn et al., 1989; Darvill et al., 1992). Oligogalacturonides, in the presence of Ca^{2+}, form intermolecular complexes named "egg boxes". Conformational analysis has established that the "egg box" conformation requires oligopectate fragments with a degree of polymerization higher than 10 (Kohn, 1975). On the basis of the correlation between the degree of polymerization required for these intermolecular conformations and elicitor activity, it has been proposed that oligogalacturonide-Ca^{2+} complexes rather than the oligogalacturonides *per se* are the active molecular signals (Liners et al., 1989; Messiaen and Van Cutsem, 1994).

Several rapid responses occurring at the plant cell surface may be part of the transduction mechanism of the oligogalacturonide signal. Oligogalacturonides have been shown to induce, within 5 min, transient stimulation of cytoplasmic Ca^{2+} influx, K^+ efflux, cytoplasmic acidification and depolarization of plasma membrane of tobacco cultured cells (Mathieu et al., 1991; Thain et al., 1995); later, they may be internalized by receptor-mediated endocytosis, as suggested for soybean suspension-cultured cells (Horn et al., 1989). The elevation of cytosolic free Ca^{2+} levels induced by oligogalacturonides has also been observed in carrot protoplasts (Messiaen et al., 1993). A role of Ca^{2+} as a second messenger in the oligogalacturonide signal transduction pathway has been proposed by Farmer et al. (1991).

Oligogalacturonides generate H_2O_2 production in cucumber (Svalheim and Robertsen, 1993), soybean (Apostol et al., 1989; Legendre et al., 1993; Chandra and Low, 1995) and castor bean (Bruce and West, 1989). Treatment of soybean cells with oligogalacturonides or with the G protein activator mastoparan determines a transient increase of IP_3, which reaches a maximum within 1 min prior to H_2O_2 production, while neomycin inhibits both IP_3 increase and the oxidative burst (Legendre et al., 1993). The oligogalacturonide-induced oxidative burst in soybean cells is inhibited by the protein kinase inhibitors K-252a and staurosporine. On the other hand, phosphatase inhibitors mimic elicitor action since they are able to induce the oxidative burst in the absence of oligogalacturonides (Chandra and Low, 1995).

Oligogalacturonides also enhance the *in vitro* phosphorylation of a 34 kD protein associated with plasma membranes of potato and tomato (pp34) (Jacinto et al., 1993; Reymond et al., 1995). In soybean cells the transduction of oligogalacturonides appears to be mediated by an unbalance in the phosphorylation/dephosphorylation equilibrium (Chandra and Low, 1995).

Oligogalacturonides of chain length varying between 10 and 15 are transiently produced by the action of the *endo*polygalacturonase on homogalacturonan. Thus, *endo*polygalacturonases release elicitor-active oligogalacturonides, but also degrade them into inactive oligomers. This implies that an extensive degradation of the cell wall homogalacturonan negatively affects the elicitor activity of the *endo*polygalacturonase. The cell wall of most plants contains a protein, PGIP (polygalacturonase-inhibiting protein), that binds to and modulates the *endo*polygalacturonase in such a way that the balance between release of elicitor-active oligogalacturonides and depolymerization of the active oligogalacturonides into inactive molecules is altered and the accumulation of elicitor-active molecules is favored (Cervone et al., 1986; Cervone et al., 1987; Cervone et al., 1989; De Lorenzo et al., 1990).

A gene encoding bean PGIP has been cloned (Toubart et al., 1992). The protein is constitutively present at low levels and can be induced in hypocotyls and suspension cultures by a variety of known elicitors, including oligogalacturonides (Bergmann et al., 1994). This suggests the occurrence of a feed-forward mechanism for the accumulation of oligogalacturonides (Cervone et al., 1993). *In situ* and Northern blot analyses in incompatible interactions between cultivars of bean and races of *Colletotrichum lindemuthianum* have shown that *pgip* transcript accumulation is rapid, transient and spatially localized in the cells at the site of infection. In contrast, the accumulation of *pgip* transcripts in compatible interactions is delayed, weaker and spatially diffuse (Nuss et al., 1996; Clark et al., unpublished results of our laboratory).

A site-directed mutagenesis approach has been utilized to study the structural basis of the elicitor activity of *endo*polygalacturonase. The sequence encoding the *endo*polygalacturonase of *Fusarium moniliforme* has been expressed in *Saccharomyces cerevisiae*. The *endo*polygalacturonase produced by the yeast possesses biochemical properties similar to the enzyme secreted by *F. moniliforme*, including the capability to elicit phytoalexin accumulation in soybean cotyledons. Replacement of His-234 with Lys abolishes enzymatic activity but, interestingly, the inactive enzyme still exhibits elicitor activity, suggesting that elicitation by *endo*polygalacturonase is not due solely to the ability of the enzyme to release oligogalacturonides from the plant cell wall (Caprari et al., unpublished results of our laboratory).

Xylanases

Fungal xylanases behave as elicitors of plant defense responses such as phytoalexin induction (Farmer and Helgeson, 1987), cell death (Bucheli et al., 1990) and ethylene production (Fuchs et al., 1989). A β-1,4-xylanase purified from *Trichoderma viride* with an "endo" mode of action induces ethylene in tobacco leaf disks. Antibodies against this enzyme, as well as heat treatment or chemical inactivation, inhibit both xylanase and ethylene-inducing activities (Dean et al., 1991). Induction of ethylene is due to the stimulation of 1-aminocyclopropane-1-carboxylic acid synthase activity (Bailey et al., 1990) and does not appear to involve release of biologically active cell wall fragments (Sharon et al., 1993). Xylanase of *T. viride* is synthesized as a 25 kD precursor and subsequently processed to a 22 kD secreted polypeptide. The enzyme has a compact conformation with an apparent Mr determined by gel permeation chromatography of about 9.2 kD, a size that is likely to allow the protein to pass through cell wall pores and interact with the plasmalemma (Dean and Anderson, 1991). Also, the *endo*xylanase of the maize pathogen *Cochliobolus carbonum* exhibits a low apparent molecular weight under non-denaturing conditions and passes through a 10 kD cut-off ultrafiltration membrane (Holden and Walton, 1992). Protoplasts are capable of responding to elicitation by xylanase, further suggesting that the protein may act directly at the level of the plasmalemma (Sharon et al., 1993). Genetic analyses of two cultivars with different sensitivities to xylanase show that elicitation of ethylene production is controlled by a single dominant gene (Bailey et al., 1993). This gene might encode the putative plant receptor of xylanase.

Xylanase from *T. viride* also elicits PR proteins in tobacco leaves through an ethylene-independent pathway (Lotan and Fluhr, 1990). In tobacco plants this enzyme induces typical defense responses such as necrosis and electrolyte leakage not only at the site of application but also at distant sites. This long-distance effect is associated with the translocation of the enzyme in the plant xylem (Bailey et al., 1991).

An interesting report by Felix et al. (1994) focuses on the importance of the phosphorylation/dephosphorylation equilibrium in the transduction of the xylanase signal. Inhibition of protein phosphatases by calyculin A determines, in suspension cultures of tomato cells, the same responses as those induced by xylanase from *T. viride*, i.e., alkalinization of the medium and the induction of 1-aminocyclopropane-1-carboxylate synthase. Both responses are inhibited by K-252a.

Lipophilic molecules

Arachidonic acid, the best characterized lipophilic molecule of fungal origin with elicitor activity, has been reported to induce phytoalexin synthesis and tissue browning in potato tuber discs (Bostock et al., 1982), rishitin and capsidiol accumulation in pepper (Bloch et al., 1984; Hoshino et al., 1994), accumulation of isoflavonoid phytoalexins in French bean (Longland et al., 1987) and accumulation of the phytoalexin lubimin in eggplant fruit discs (Castoria et al., 1995). Also, some esterified forms of arachidonic acid such as phosphatidylcholine, β-arachidonoyl, γ-stearoyl or short arachidonic acid acyl esters exhibit elicitor activity (Preisig and Kuc, 1985; Fanelli et al., 1992).

In potato tuber tissue arachidonic acid induces the expression of the *hmg2* gene, encoding an isoform of 3-hydroxy-3-methylglutaryl coenzyme A reductase involved in phytoalexin synthesis. At the same time arachidonic acid suppresses the expression of the *hmg1* gene, encoding another isoform of the same enzyme involved in the steroid glycoalkaloid synthesis. Staurosporine blocks the arachidonic acid-induced expression of *hmg2* and consequently phytoalexin accumulation. On the other hand, the same inhibitor prevents the suppression played by arachidonic acid on the *hmg1* gene expression, suggesting that staurosporine-sensitive protein kinase(s) plays a role both in the positive regulation of the expression of *hmg2* and in the negative regulation the expression of *hmg1* (Choi and Bostock, 1994).

A linkage between the elicitor activity of arachidonic acid and its metabolization by lipoxy-genase (LOX) has been suggested (Vaughn and Lulai, 1992; Choi and Bostock, 1994). For example, 5-S-HPETE elicits both phytoalexin synthesis and tissue browning (Castoria et al., 1992). In contrast, neither induction of phytoalexin synthesis nor hypersensitive tissue browning by the products of LOX activity on arachidonic acid have been observed by other authors (Ricker and Bostock, 1994).

While arachidonic acid is perceived by plant cells in the micromolar range, ergosterol, the main sterol in most higher fungi, is perceived in the subnanomolar range in tomato cells (Granado et al., 1995).

Glucan

Glucan fragments deriving from the cell wall of many fungi are potent elicitors of plant defense responses. Studies on the possible mechanisms for the transmission of the glucan signal within the plants cells have been recently reviewed by Côté and Hahn (1994).

The smallest fragment with strong elicitor activity, purified from a complex mixture of oligoglucosides released from the mycelial walls of *Phytophthora sojae,* corresponded to a hepta β-glucoside (Sharp et al., 1984). Structure-function analysis elucidated the elements of the hepta β-glucoside elicitor which are important for its biological activity: all three non-reducing terminal glucosyl residues and their distribution along the backbone of the molecule are critical for the elicitor activity in the soybean bioassay (Cheong et al., 1991).

The first evidence of a glucan receptor that mediates phytoalexin accumulation in soybean was obtained by Yoshikawa et al. (1983) using [^{14}C]-mycolaminaran as a radiolabeled ligand. Unlabeled ligand competed for binding sites but a large amount of radioactivity was not displaced. Maximum binding was achieved after 30 min and occurred at pH between 7 and 8.

Specific binding to plasma membrane fractions of soybean was reported for a mixture of β-1,3 [^{3}H]-glucans obtained by acid hydrolysis of *Phytophthora megasperma* cell walls and tritiation by reduction with sodium boro[^{3}H]-hydride. Only polysaccharides of a branched β-glucan type were effective in competition studies (Cosio et al., 1988; Schmidt and Ebel, 1987). Subsequently, binding sites specific for the homogeneous hepta-β-glucoside were found in soybean plasma membranes. Most importantly, competitive inhibition of binding of the radiolabelled elicitor by a number of structurally related oligoglucosides demonstrated a direct correlation between binding affinities and the elicitor activities of these oligoglucosides (Cosio et al., 1990b; Cheong and Hahn, 1991).

Solubilization from soybean plasma membranes of the glucan binding proteins by detergents allowed partial purification of the putative receptor(s) (Cosio et al., 1990a; Cheong et al., 1993). Most elicitor binding activity was associated with large detergent-protein micelles. Disruption of the micelles using high detergent concentrations and sonication resulted in the loss of 70% binding activity suggesting that elicitor binding proteins exist in a multimeric complex. Photoaffinity labelling resolved three proteins of Mr 70 000, 100 000 and 170 000; the latter two proteins were observed only after extended exposure times (Cosio et al., 1992). A glucan affinity matrix composed of a branched (1→3, 1→6)-linked β-glucan fraction allowed the identification of three proteins with Mr of 70 000, 97 000 and 33 000 (Frey et al., 1993). It is unclear whether the 97 000 and 100 000 proteins are identical. Determining whether or not all the proteins identified by an affinity purification or photoaffinity labeling are essential for elicitor-binding activity will require further purification and characterization (Côté et al., 1995).

Chitin and chitosan

Chitin, a major cell wall component of fungi, is an elicitor of plant defense responses (Barber et al., 1989; Roby et al., 1987; Ren and West, 1992). Treatment of rice suspension-cultured cells with chitin increases the amount of chitinase activity detected in the culture filtrate (Ren and West, 1992). N-acetylchitooctaose conjugated with tyramine by reductive amination and labeled with ^{125}I shows elicitor activity and binds specifically and reversibly to rice cell membranes with a Kd of 5.4 nM. The estimated concentration of binding sites is 0.3 pmol/mg protein. Competitive inhibition using N-acetylchitoheptaose, another elicitor-active chitin fragment, is almost complete (Shibuya et al., 1993).

Several biological responses are induced by chitin binding in tomato cells, including the induction of extracellular alkalinization and changes in protein phosphorylation; these effects may be part of the signal transduction cascade leading to induction of defense responses. The induction of extracellular alkalinization cells by oligochitins is transient and is followed by a refractory state where the cells are no longer responsive to subsequent treament with oligochitin, but they still respond to preparations of fungal xylanase (Felix et al., 1993). Only chitin fragments with a DP ≥ 4 show elicitor activity.

Specific binding of chitin fragments to tomato cell membranes and protoplasts has been reported: a chitin fragment with DP 5, aminated at the reducing end and coupled to t-butoxycarbonyl-L-[^{35}S]methionine *via* an amidoglycine spacer, specifically, saturably and reversibly binds to whole cells and microsomal membrane fractions (Kd 1.43 and 23 nM, respectively). Binding is competed by chitin fragments of different lengths; the relative efficiency of chitin fragments of different sizes in binding competition closely corresponds to their relative efficiency in inducing the alkalinization response in tomato cells. Binding is also competed by derivatives of chitooligosaccharides and by Nod factors (Baureithel et al., 1994). Nod factors are lipo-chitooligosaccharides secreted by the bacterial genera *Rhizobium* and *Bradyrhizobium* which induce the formation of nodules on leguminosae, their host plants; they have been shown to induce the alkalinization response in tomato cells (Staehelin et al., 1994). The activity of the Nod factors as competitors of binding and as inducers of the alkalinization response is about one order of magnitude lower than that of a chitin fragment with DP of 4 (Baureithel et al., 1994).

Chitosan, a β-1,4-D-glucosamine derivative of chitin also found in the walls of several fungi, elicits the accumulation of the phytoalexin pisatin in pea (Kendra and Hadwiger, 1984), and production of protease inhibitors in tomato leaves (Walker-Simmons et al., 1983). Characterization of molecules with various degrees of polymerization has demonstrated that although high molecular weight chitosan fragments (heptamer units) are most active in inducing phytoalexin accumulation, trimer units at a higher concentration are also active. The octadecanoid pathway has been implicated in the transmission of the chitosan signal leading to the accumulation of protease inhibitors in tomato (Doares et al., 1995).

Elicitins

Elicitins are a small family of highly conserved 10-kD proteins secreted by various species of *Phytophthora* (Nespoulous et al., 1992). Two classes of elicitins exist according to toxicity: the acidic α-elicitins are less toxic and include capsicein and parasiticein while the basic β-elicitins are more toxic and include cryptogein and cinnamomin. Application of purified elicitins to tobacco can cause hypersensitive cell death and subsequent increased resistance to fungal infection (Yu, 1995). Tobacco suspension-cultured cells treated with cryptogein generate active oxygen species and undergo alkalization of the medium (Viard et al., 1994). Treatment of tobacco cells with elicitins also activates the transcription of the retrotransposon Tnt1 (Pouteau et al., 1994).

Not all species respond to infiltration with elicitins. Purified elicitins from *P. parasitica* and *P. cryptogea* induce hypersensitive responses in *Nicotiana* species and some radish and turnip cultivars, but not in 12 other plant species (Kamoun et al., 1993). In another study elicitin-induced necrosis was found on all the solanaceous plants tested. Nevertheless, some cultivars within the responsive species were unresponsive, indicating a possible gene-for-gene interaction (Pernollet et al., 1993).

Protein phosphorylation is likely to occur after the addition of cryptogein to tobacco cells. Simultaneous addition of cryptogein and staurosporine abolishes medium alkalinization; if staurosporine is added after medium alkalinization, the effect of cryptogein is reversed. In tobacco cells staurosporine prevents the cryptogein-induced oxidative burst, alkalinization of the medium and K^+ efflux (Viard et al., 1994).

Binding studies performed with ^{125}I-labeled cryptogein suggest the presence of a single class of binding sites (K_d of 2 nM) (Wendehenne et al., 1995).

A glycopeptide from yeast

A glycopeptide fraction obtained by enzymatic hydrolysis of yeast invertase elicits ethylene production in tomato cells. The fraction has been separated into five glycopeptides having the same amino acid sequences and differing N-glycan moieties. The number of mannosyl residues affects the activity of these molecules, with the most active containing 10–12 residues. Treatment

of the elicitors with the *endo*glycosidase Endo H abolishes activity, and the carbohydrates generated by this procedure suppress the elicitor activity of the glycopeptides by competitive inhibition
(Basse et al., 1992).

A single class of high affinity binding sites with an apparent K_d of 3.3 nM is detected in membranes prepared from tomato suspension-cultured cells. Glycopeptides with N-linked side chains
of 9–11 mannosyl residues have a high affinity for the binding sites as do specific glycan
suppressors (Basse et al., 1993).

A peptide from Phytophthora megasperma

A 42 kD glycoprotein purified from *P. megasperma* f. sp. *glycinea* elicits phytoalexin accumulation in parsley cells. A peptide of 13 amino acids (Pep-13), identified within the intact glycoprotein, is sufficient to stimulate the same responses as the entire glycoprotein elicitor. Pep-13
has been labeled at the Tyr-12 residue with ^{125}I and used in binding studies with either parsley
protoplasts or microsomal membranes. The existence at the level of the plasma membrane of a
high affinity, single class of binding sites with a Kd of 2.4 nM has been demonstrated. The
apparent concentration of binding sites is 88 fmol/mg protein which represents an average of
1600 binding sites per cell. Competitive binding and elicitation experiments with derivatives of
Pep-13 have established a functional link between binding activity and biological response.
Elimination of either one of the aromatic amino acids Trp-2 or Tyr-12 leads to a considerable
reduction of the elicitor activity. Identification of the importance of the individual amino acids has
been achieved using a series of Ala substitutions. Replacement of Trp-2 or Pro-5 severely
reduces the ability of these analogs to elicit plant responses and renders them inactive as
competitors for binding (Nürnberger et al., 1994; Diekmann et al., 1994). Chemical crosslinking
of ^{125}I-Pep-13 to parsley microsomal membranes has been achieved and subsequent solubilization and separation of the membrane proteins allowed the identification of a 91 kD protein. The
competitors used in the binding studies were able to inhibit the cross-linking of the ^{125}I-Pep-13
to parsley microsomal membranes. Attempted crosslinking of the ligand to membranes of other
species (carrot, soybean, *Arabidopsis*) was not successful, indicating a possible species
specificity of Pep-13 (Nürnberger et al., 1995).

The phytoalexin synthesis induced by the 42 kD protein is prevented by the inhibitor of phosphatases, okadaic acid, but not by the kinase inhibitors K-252a and staurosporine (Renelt et al.,
1993). Pep-13 stimulates Ca^{2+} and H^+ influx, as well as an efflux of K^+ and Cl^- ions, an oxidative burst and accumulation of jasmonate in parsley cells. Inhibition of the ion fluxes, which are
the earliest reactions to elicitor treatment, prevents the activation of all other responses, including

phytoalexin synthesis. *Vice versa*, stimulation of the ion fluxes in the absence of elicitor, using the polyene antibiotic amphotericin B, initiates defense gene activation, phytoalexin production and an oxidative burst. Inhibition of lipoxygenase blocks the elicitor-stimulated synthesis of jasmonate, but does not affect defense gene activation and phytoalexin accumulation (Hahlbrock et al., 1995).

A glycopeptide from Puccinia graminis

A glycoprotein elicitor purified from cell walls of *P. graminis* f. sp. *tritici* induces lignification in wheat leaves. The elicitor has been labelled and high affinity binding (Kd 2 μM) to two membrane proteins (Mr 30000 and 34000) has been demonstrated (Kogel et al., 1991).

The AVR products of Cladosporium fulvum

C. fulvum, a fungal pathogen of tomato, occurs in several races, each of which is able to attack different tomato cultivars. The interaction between *C. fulvum* and tomato is a well-established model system for studying the gene-for-gene relationships since each characterized race of the fungus carries a specific gene (*Avr* gene) which confers to that race the capability to elicit HR in those tomato cultivars carrying the corresponding resistance genes (*R* genes). Several *Avr* genes have been identified in *C. fulvum,* some of which have been cloned and characterized.

The A*vr9* gene encodes a precursor protein of 63 amino acids subsequently processed into a 28-amino acid peptide elicitor (van den Ackerveken et al., 1993). The mature 28-amino acid elicitor induces HR only in those tomato cultivars carrying the *Cf-9* resistance gene. Apparently, the races that are virulent on tomato *Cf-9* genotypes have no DNA homologous to the coding sequence of *Avr9*. If a strain of the fungus naturally virulent towards *Cf-9* tomatoes is transformed with *Avr9,* it becomes avirulent on *Cf-9* tomatoes (van den Ackerveken et al., 1992). *Cf-9,* a gene which has been cloned recently (Jones et al., 1994), might encode a receptor protein for this peptide. Binding studies are in progress to establish whether the products of the *Avr9* and *Cf-9* genes specifically interact.

The A*vr4* gene of *C. fulvum* has also been cloned and characterized; it encodes a pre-protein of 135 amino acids which is processed into a 117-amino acid mature protein. Races of *C. fulvum* carrying *Avr4* are avirulent on tomato cultivars carrying the resistance gene *Cf-4*. It has been shown that a single base-pair change in *Avr4* leads to virulence of races previously avirulent on *Cf-4* tomatoes. Here again the *Cf-4* product is believed to be the receptor for the *Avr4* product.

Homologues of the *Avr4* gene have been isolated and sequenced from different races of *C. fulvum* virulent on tomato *Cf-4* genotypes. Every *Avr4* homologue contains a single base pair mutation in one of the three different cysteine codons (Joosten et al., 1994). The fact that the the products of the genes *Avr4*, *Avr9*, *Nip-1* (see later) as well as elicitins are all cysteine-rich peptides points to the importance of cystine residues in the specificity and pathogenicity of fungi (Templeton et al., 1994).

The product of the Nip-1 *gene of* Rhynchosporium secalis

The *R. secalis*-barley interaction also conforms to the gene-for-gene hypothesis. The avirulence gene *Nip1* has been identified and cloned. The gene product NIP1 is a small phytotoxic protein which behaves as a race-specific elicitor of defense responses in barley cultivars carrying the resistance gene *Rrs1*. All fungal races that are avirulent on barley cultivars of the *Rrs1* resistance genotype carry and express the *Nip1* gene. Transformation of a virulent race with the *Nip1* gene yields an avirulent transformant confirming the status of avirulence gene. A race which carries the *Nip1* gene with a single nucleotide exchange in the coding region is virulent on *Rrs1* plants as the mutated NIP1 product is inactive as elicitor on these plants (Rohe et al., 1995).

Other studies on the transduction of fungal elicitors

In addition to the studies carried out with purified elicitors, many attempts to clarify the possible transduction pathways have been performed by using crude or partially purified preparations of fungal origin. As in numerous other fields of plant biology, the starting point for most of these studies has been the search for similarities with the animal signal transduction pathways. Because of the multiplicity of the different elicitors used, these studies will be described focusing on the involvement of each single transduction component, rather than on each single elicitor preparation. Several reviews have been recently published on this topic (Dixon et al., 1994; Côté and Hahn, 1994; Ebel and Cosio, 1994; Boller, 1995).

Ion fluxes

An immediate four-fold increase in the plasma membrane H^+-ATPase activity of a resistant tomato cell line is determined by an elicitor preparation from an avirulent race of *C. fulvum*. Such an increase appears to be regulated by elicitor-induced dephosphorylation of the enzyme, likely to be mediated by G proteins. Immunodetection of *in vitro* labeled [γ-^{32}P]ATP plasma membranes with an antibody raised against the C-terminal H$^+$-ATPase of *Avena sativa* shows the dephosphorylation of a band corresponding to the 100 kD ATPase subunit (Vera-Estrella et al., 1994a).

Ca^{2+}

Several studies indicate a calcium requirement for induction of defense responses by fungal elicitors. The absence of Ca^{2+} suppresses the production of active oxygen species upon treatment with an elicitor preparation from the cell wall of *Heterobasidion annosum* in spruce cultured cells, whereas the Ca^{2+} ionophore A23187 alone induces an oxidative burst, although to a lower degree with respect to the elicitor (Schwacke and Hager, 1992). The Ca^{2+} channel blocker verapamil delays the phytoalexin accumulation induced in soybean by a crude glucan elicitor preparation from *P. megasperma* f. sp. *glycinea* (Stäb and Ebel, 1987). Verapamil also inhibits the elicitation of phytoalexin synthesis induced by the Ca^{2+} ionophore A23187 and cAMP in carrot cells (Kurosaki et al., 1987). Strong evidence for the role of Ca^{2+} in the elicitor transduction pathway has been provided by the use of transgenic plants containing the Ca^{2+}-sensitive protein aequorin. In these plants a rapid and transient increase of intracellular Ca^{2+} concentration occurs upon treatment with elicitor preparations from yeast and *Gliocladium deliquescens* (Knight et al., 1991). Conflicting results have been obtained on the possible involvement of calmodulin in the elicitation process. Antagonists of calmodulin prevent phytoalexin synthesis in tobacco cells treated with a crude cell wall preparation from *P. parasitica* (Vögeli et al., 1992), but have no effect in soybean (Stäb and Ebel, 1987). It may be relevant that a Ca^{2+}-dependent protein kinase from soybean contains a calmodulin-like regulatory domain (Harper et al., 1991).

cAMP

cAMP levels rise significantly in bean cells 15 min after treatment with elicitor preparations from *P. megasperma* f. sp. *glycinea*, *Botrytis cynerea* and *Colletotrichum lindemuthianum* (Bolwell, 1992). Challenge of *Medicago sativa* with a glycoprotein elicitor from *Verticillium albo-atrum* determines an approximately four-fold transient increase of cAMP and adenyl cyclase activity levels within 4 min, followed by an increase in cAMP phosphodiesterase activity (Cooke et al., 1994). Dibutyryl cAMP, a derivative of cAMP, is an elicitor of the phytoalexin medicarpin in

Medicago (Cooke et al., 1994) as well as – in the presence of Ca^{2+} – of the phytoalexin 6-methoxymellein in carrot cells (Kurosaki et al., 1987).

Phosphoinositides

The number of reports on the involvement of the IP_3 signaling system (or some of its components) in elicitor transduction has been steadily increasing in the past few years. In parsley cells and in pea treated with a crude preparation from hyphal cell wall of *P. megasperma* f. sp. *glycinea* and with an elicitor from *Mycosphaerella pinodes*, respectively, the accumulation of the phytoalexins is prevented by the inhibitor of phospholipase C, neomycin (Renelt et al., 1993; Toyoda et al., 1993). Increases of PIP_2 and IP_3 have been observed in pea epicotyls within 15 min of elicitor treatment. Treatment of isolated plasma membranes of pea with elicitors induces a rapid incorporation of ^{32}P-ATP into phosphatidylinositol 4-phosphate (PIP) and phosphatidylinositol 4-5 bisphosphate (PIP_2) (Toyoda et al., 1993).

Recent findings suggest a role for inositol 1-4 bisphosphate (IP_2) rather than for IP_3. Treatment of tobacco cells with a water soluble elicitor from *Phytophthora nicotianae* var. *nicotianae* induces a phospholipid turnover which precedes the induction of PAL activity. Ten minutes after elicitor treatment the level of IP_2 is 15 fold higher than the control, whereas only a 1.38-fold increase is observed for IP_3 (Kamada and Muto, 1994).

G proteins

The possible role of G proteins in the activation of phosphatase(s) has been proposed for the stimulation of H^+ ATPase in isolated plasma membranes of a resistant tomato cell line in response to an elicitor preparation from an avirulent race of *C. fulvum*. The guanidine nucleotide analog GTP(γ)S and mastoparan determine increases of ATPase activity similar to that induced by the elicitor. The same compounds behave as elicitors in stimulating the putative active oxygen generating system of the plasma membrane, and in deactivating (or suppressing) the scavenger activity of ascorbate peroxidase (Vera-Estrella et al., 1994b).

Protein phosphorylation and dephosphorylation

The phosphorylation of a set of proteins has been described in suspension-cultured parsley cells treated with an elicitor preparation from *P. megasperma* f. sp. *glycinea*. A 45 kD protein associated with cell membranes is phosphorylated after only 1 min upon treatment. The phosphorylation of a 26 kD nuclear protein is also observed after a few minutes. The phosphorylation events are temporally related to the activation of genes encoding enzymes of the phytoalexin synthesis pathway. Moreover, phosphorylation is Ca^{2+}-dependent and elicitor-specific, since it does not occur following the exposure of parsley cells to other stress factors such as UV irradiation, heat shock or treatment with mercuric chloride (Dietrich et al., 1990).

The protein kinase inhibitors K-252a and staurosporine block ethylene production, PAL activity and alkalinization of the medium in suspension-cultured tomato cells treated with fungal elicitors (Grosskopf et al., 1990; Felix et al., 1991), and the oxidative burst induced by an elicitor preparation from *P. megasperma* f. sp. *glycinea* in soybean cells (Levine et al., 1994). This last effect is also inhibited by the protein phosphatase 2A inhibitor cantharidin, but not by cypermethrin, a protein phosphatase 2B inhibitor. Staurosporine blocks the oxidative burst induced in spruce cells by a cell wall elicitor from *H. annosum* (Schwacke and Hager, 1992).

A rapid and transient induction of a 47 kD protein kinase that is active on serine and threonine residues, as well as the phosphorylation of a tyrosine residue of the 47 kD kinase itself, occurs in tobacco cells treated with an elicitor derived from the cell walls of *Phytophthora infestans*. Both staurosporine and Gd^{3+}, a blocker of Ca^{2+} channels, inhibit the elicitor-induced kinase activation; staurosporine also inhibits the activity of the enzyme (Suzuki and Shinshi, 1995).

Jasmonic acid

Much attention has recently been paid to the role of jasmonic acid (JA) and its methyl esters (MeJA), two prostaglandin-like metabolites that are synthesized by the action of LOX(s) on α-linolenic acid followed by hydroperoxide dehydrase and β-oxidation reactions. These two compounds have been proposed as second messengers in wound-induced responses (Farmer and Ryan, 1992). The possibility that JA and its derivatives are an integral part of the signal transduction pathway linking the elicitor-receptor interaction with the activation of defense-related genes is emerging in a number of studies (Mueller et al., 1993; Gundlach et al., 1992; Kauss et al., 1994; Doares et al., 1995).

Active oxygen species and lipid peroxidation

In plants, the trigger of a rapid oxidative burst, in which activated oxygen species (AOS) such as superoxide anion and hydrogen peroxide are formed within a few minutes, has been observed upon treament with elicitors. Production of AOS and processes associated with this production, such as the oxidative cross-linking of the cell wall and lipid peroxidation, have been proposed to contribute to both programmed cell death and rapid activation of defense responses (Levine et al., 1994). Generation of hydrogen peroxide as well as other AOS may actually be a defense response *per se*. However, since the oxidative burst takes place prior to the activation of transcription-dependent defenses, it is also possible that AOS function as second messengers in the elicitor transduction pathway. For example, in soybean, H_2O_2 induces an accumulation of mRNA encoding glutathione-S-transferase (GST) that is comparable to that induced by the glucan elicitor from *P. megasperma* (Levine et al., 1994; Tenhaken et al., 1995).

As in mammalian inflammatory responses, AOS might participate in redox signaling and in the generation of bioactive fatty acid-derivatives analogous to prostaglandins. Immune responses in animal cells are known to be mediated by the redox-sensitive *trans*-factor NF-κB (Schreck et al., 1991). Recently, the activation of a redox-regulated *trans*-factor has been reported also in elicited plant cells; it binds to a specific 3' flanking region of a bean mRNA, coding for a proline-rich protein, which undergoes a significant decrease of its half-life upon elicitor treatment (Zhang et al., 1993; Zhang and Mehdy, 1994).

A rapid lipid peroxidation has been observed in plant cells treated with fungal-derived elicitors, but the possible signaling activities of the products generated by the peroxidative process have not been clarified yet (Chai and Doke, 1987; Rogers et al., 1988).

LRR modular proteins in the recognition of fungal elicitors

The nature of the receptors of the fungal elicitors so far characterized is still elusive. Characterization of the receptor proteins as well as cloning of the corresponding genes is awaited to answer important questions about their recognition mechanisms. Fungal molecules as different as those we have considered in this chapter are recognized by plants; do correspondingly many different classes of receptors exist in the plant? The recent cloning of several plant resistance genes is shedding some light on the nature of some of the elements involved in the perception and/or transduction of microbial elicitors such as the products of *Avr* genes. The analysis of the R genes raises the hypothesis that most of the plant receptors specialized for the recognition of non-self molecules may share common structural characteristics: most of the R genes characterized encode leucine-rich repeat (LRR) proteins. Therefore, the LRR protein structure may have been selected, during evolution, for the accomplishment of the plant immunological functions. It would not be a surprise if some receptors for the elicitors discussed in this chapter turned out to be LRR proteins.

LRR proteins

In organisms other than plants, from bacteria to humans, proteins containing LRR appear to be specialized for protein-protein interactions. Many of them also interact with membranes and possess domains for signal transduction (Kobe and Deisenhofer, 1994). Over 30 LRR proteins have been characterized: among these, the human LRG, a putative membrane-derived serum protein; the regulatory domains or subunits of the enzymes adenylyl cyclase and carboxypeptidase N; the cell adhesion protein *Toll* and the photoreceptor membrane glycoprotein chaoptin of *Drosophila*; the human extracellular matrix binding glycoprotein decorin; the alpha chain of human platelet glycoprotein (GpIb), a transmembrane receptor; receptors for gonadotrophins and neurotrophins; the *sds22+* nuclear protein; and a porcine RNAse inhibitor (Kobe and Deisenhofer, 1994 and references therein). This last protein exhibits a non-globular structure constructed of alternating β-sheets and α-helixes (Kobe and Deisenhofer, 1995).

The structural similarity between the repeat regions of these proteins, in spite of the diversity of their functions, may indicate an evolutionary conservation between the proteins, or, more likely, reflect the coincidental and convergent evolution of a protein domain, with a strong selection pressure mantaining this structure. The common feature of the LRR proteins, i.e. the ability to bind other proteins or membranes, suggests a specialization of LRR domains for the establishment of strong interactions between macromolecules.

The first LRR protein characterized in plants was PGIP (Toubart et al., 1992). The similarity between PGIP and the putative extracellular receptor-like domain of a cloned *A. thaliana* receptor-like protein kinase (RLK5) (Walker, 1993) is of particular interest. The catalytic domain of RLK5 is homologous to that present in the deduced protein of other cloned genes (*ZmPK1, RLK1, RLK4*) (Walker, 1993) which instead exhibit putative extracellular domains related to the products of *Brassica S* Locus-related *SLG* and *SRK* genes (Nasrallah and Nasrallah, 1993; Nasrallah et al., 1994). SLG (S Locus Glycoprotein) is a secreted glycoprotein, while SRK (S Locus Receptor Kinase) is a transmembrane receptor protein kinase; SLG and the extracellular receptor domain of SRK isolated from the same S haplotype share a high level (>80%) of sequence identity. A model has been proposed in which SRK, acting in combination with SLG, couples the recognition event at the pollen-stigma interface to a cytoplasmic phosphorylation cascade that leads to pollen rejection and therefore to self-incompatibility (Nasrallah et al., 1994; Boyes and Nasrallah, 1995)

The analogy between the recognition system involved in pollen-stigma interactions and that involved in plant-pathogen interactions at the cellular and genetic levels, the similarity between PGIP and the extracellular domain of a receptor-like protein kinase RLK5, and the ability of

PGIP to recognize a fungal molecule, raised the hypothesis that PGIP and PGIP-like proteins may act as secreted "receptors" in different aspects of plant-fungus recognition, such as those leading to non-host resistance or race-cultivar specificity. Two-component (secreted receptor/transmembrane receptor-kinase) signaling systems, similar to that controlling self-incompatibility in *Brassica* but differing for the LRR PGIP-like receptor domains and perhaps the type of kinase domain, may be present on the plant cell surface (De Lorenzo et al., 1994).

In the attempt to elucidate the role of PGIP in plant resistance to fungi, different *pgip*-related genes are being characterized. It has been shown that a family of *pgip* genes, likely to be clustered on chromosome 10, is present in the genome of *Phaseolus vulgaris* (Frediani et al., 1993) and several *pgip*-related clones have already been isolated in our laboratory. The structural and functional analyses of these clones are in progress. Also, the possible presence of membrane-bound PGIP-related proteins is being investigated. In this connection, it is of interest that *endo*polygalacturonase of *C. lindemuthianum* has been shown to interact with protoplasts of *P. vulgaris*, suggesting the presence of a polygalacturonase-binding protein (a PGIP?) at the level of plasmalemma (Cervone and De Lorenzo, 1985).

The products of the R genes

After decades of intense research seven plant genes (R) which participate in gene-for-gene resistance have been cloned. Five of these genes code for LRR proteins (the gene *N* of tobacco for resistance to tobacco mosaic virus, the *RPS2* and *RPM1* genes of *Arabidopsis* for resistance to two different pathovars of the bacterium *Pseudomonas syringae*, the gene *Cf-9* of tomato for resistance to *C. fulvum*, the gene *Xa21* of rice for resistance to *Xanthomonas oryzae* pv. *oryzae* race 6) (Dangl, 1995; de Wit, 1995; Song et al., 1995; Staskawicz et al., 1995, and references therein). Another gene (L^6 of flax) encodes a leucine-rich protein related to the products of *RPS2* and *N*, but without an obvious repeated structure (Lawrence et al., 1995). The tomato *Pto* gene codes for a kinase (Martin et al., 1993), and, to confer resistance, requires the presence of the gene *Prf*, coding an LRR protein homologous to the product of *RPS2* (Staskawicz et al., 1995). The proteins encoded by the genes *N*, *RPS2*, L^6, *RPM1* and *Prf* have as a common characteristic, in addition to the LRR structure, the presence of a putative nucleotide binding site (P loop), similar to one found in Ras proteins and in the β subunit of the ATP synthase. The protein encoded by the tomato *Cf-9* gene (Jones et al., 1994) has a putative extracellular LRR N-terminal domain related to PGIP and a hydrophobic putative transmembrane domain at the C-terminus. The protein encoded by the rice *Xa21* gene carries a LRR motive related to PGIP, a hydrophobic putative transmembrane domain and a serine-threonine kinase-like domain (Song et al., 1995).

For none of these gene products has the interaction with the corresponding pathogen-derived molecules yet been demonstrated.

Conclusions

The recent cloning of several resistance genes is shedding some light on the possible mechanisms of recognition and transduction of some elicitors, products of the *Avr* genes. Most of the isolated resistance genes encode proteins which share the common characteristic of a leucine-rich repeat (LRR) structure. In plants, the LRR protein structure seems therefore to have been selected during evolution for recognition of non-self molecules in race-cultivar interactions, and possibly in non-host interactions.

A widespread cell wall protein which recognizes and binds fungal *endo*polygalacturonases, the PGIP, is an LRR protein. PGIP may therefore belong to a super-family of proteins, which includes the resistance gene products, specialized for recognition of non-self molecules and rejection of pathogens. The interaction between PGIP and fungal *endo*polygalacturonases may function in a perception mechanism leading to incompatibility: PGIP may interact with membrane anchored PGIP-like proteins to transmit information through the plasma membrane (De Lorenzo et al., 1994). It is an open question whether PGIPs and PGIP-like proteins may interact with molecules other than *endo*polygalacturonases.

The *endo*polygalacturonase-PGIP recognition offers the opportunity to study the interaction between LRR proteins and their ligands at the molecular level. A family of *pgip* genes and, likely, a family of PGIP proteins, is present in *P. vulgaris*; the different PGIPs may differ in specificities and expression patterns. It is under investigation how many PGIP proteins are expressed, under what circumstances they are expressed, and whether they possess different specificities.

Many fungal molecules acting as non (race-cultivar) specific elicitors have been identified and for many of them the signal transduction pathways leading to the activation of the defense responses are being elucidated. The elucidation of the nature of their receptors, as well as the characterization and cloning of the corresponding genes, is awaited to know whether they share common characteristics with the known resistance gene products: are they also LRR proteins?

Acknowledgements
Work in the authors' laboratory is supported by the Ministero delle Risorse Agroindustriali e Forestali (MIRAAF) and by the European Community Grants CHRX-CT93-0244 and AIR 3-CT94-2215. We thank Dr. T. Boller for critically reading the manuscript.

References

Anderson, A.J. (1989) The biology of glycoproteins as elicitors. *In*: T. Kosuge and E. Nester (eds): *Plant-Microbe Interactions. Molecular and Genetic Perspectives,* Vol. 3. McGraw Hill Inc. New York, NY, pp 87–130.

Apostol, I., Heinstein, P.F. and Low, P.S. (1989) Rapid stimulation of an oxidative burst during elicitation of cultured plant cells. Role in defense and signal transduction. *Plant Physiol.* 90: 109–116.

Bailey, B.A., Dean, J.F.D. and Anderson, J.D. (1990) An ethylene biosynthesis-inducing endoxylanase elicits electrolyte leakage and necrosis in *Nicotiana tabacum* cv Xanthi leaves. *Plant Physiol.* 94: 1849–1854.

Bailey, B.A., Taylor, R., Dean, J.F.D. and Anderson, J.D. (1991) Ethylene biosynthesis-inducing endoxylanase is translocated through the xylem of *Nicotiana tabacum* cv Xanthi plants. *Plant Physiol.* 97: 1181–1186.

Bailey, B.A., Korcak, R.F. and Anderson, J.D. (1993) Sensitivity to an ethylene biosynthesis-inducing endoxylanase in *Nicotiana tabacum* L. cv Xanthi is controlled by a single dominant gene. *Plant Physiol.* 101: 1081–1088.

Barber, M.S., Bertram, R.E. and Ride, J.P. (1989) Chitin oligosaccharides elicit lignification in wounded wheat leaves. *Physiol. Molec. Plant Pathol.* 34: 3–12.

Basse, C.W., Bock, K. and Boller, T. (1992) Elicitors and suppressors of the defense response in tomato cells. Purification and characterization of glycopeptide elicitors and glycan suppressors generated by enzymatic cleavage of yeast invertase. *J. Biol. Chem.* 267: 10258–10265.

Basse, C.W., Fath, A. and Boller, T. (1993) High-affinity binding of glycopeptide elicitor to tomato cells and microsomal membranes and displacement by specific glycan suppressors. *J. Biol. Chem.* 268: 14724–14731.

Baureithel, K., Felix, G. and Boller, T. (1994) Specific, high affinity binding of chitin fragments to tomato cells and membranes. Competitive inhibition of binding by derivatives of chitooligosaccharides and a Nod factor of *Rhizobium. J. Biol. Chem.* 269: 17931–17938.

Bergmann, C., Ito, Y., Singer, D., Albersheim, P., Darvill, A.G., Benhamou, N., Nuss, L., Salvi, G., Cervone, F. and De Lorenzo, G. (1994) Polygalacturonase-inhibiting protein accumulates in *Phaseolus vulgaris* L. in response to wounding, elicitors, and fungal infection. *Plant J.* 5: 625–634.

Bloch, C.B., de Wit, P.J.G.M. and Kuc, J. (1984) Elicitation of phytoalexins by arachidonic and eicosapentaenoic acids: a host survey. *Physiol. Plant Pathol.* 25: 199–208.

Boller, T. (1995) Chemoperception of microbial signals in plant cells. *Annu. Rev. Plant Physiol.* 46: 189–214.

Bolwell, G.P. (1992) A role for the phosphorylation in the down-regulation of phenylalanine ammonia-lyase in suspension-cultured cells of French bean. *Phytochemistry* 31: 4081–4086.

Bostock, R.M., Laine, R.A. and Kuc, J.A. (1982) Factors affecting the elicitation of sesquiterpenoid phytoalexin accumulation by eicosapentaenoic and arachidonic acids in potato. *Plant Physiol.* 70: 1417–1424.

Boyes, D.C. and Nasrallah, J.B. (1995) An anther-specific gene encoded by an S locus haplotype of *Brassica* produces complementary and differentially regulated transcripts. *Plant Cell* 7: 1283–1294.

Broekaert, W.F. and Peumans, W.J. (1988) Pectic polysaccharides elicit chitinase accumulation in tobacco. *Physiol. Plant.* 74: 740–744.

Bruce, R.J. and West, C.A. (1989) Elicitation of lignin biosynthesis and isoperoxidase activity by pectic fragments in suspension-cultures of castor bean. *Plant Physiol.* 91: 889–897.

Bucheli, P., Doares, S.H., Albersheim, P. and Darvill, A. (1990) Host-pathogen interactions XXXVI. Partial purification and characterization of heat-labile molecules secreted by the rice blast pathogen that solubilize plant cell wall fragments that kill plant cells. *Physiol. Molec. Plant Pathol.* 36: 159–173.

Castoria, R., Fanelli, C., Fabbri, A.A. and Passi, S. (1992) Metabolism of arachidonic acid involved in its eliciting activity in potato tuber. *Physiol. Molec. Plant Pathol.* 41: 127–137.

Castoria, R., Fanelli, C., Zoina, A. and Scala, F. (1995) Analysis of fatty acids in lipids of *Verticillium dahliae* and induction of lubimin in eggplant. *Plant Pathol.* 44: 791–795.

Cervone, F. and De Lorenzo, G. (1985) Pectic enzymes as phytotoxins: absorption of polygalacturonase from *Colletotrichum lindemuthianum* to French bean protoplasts. *Phytopath. Med.* 24: 322–324.

Cervone, F., De Lorenzo, G., Degrà, L. and Salvi, G. (1986) Interaction of fungal polygalacturonase with plant proteins in relation to specificity and regulation of plant defense response. *In*: B. Lugtenberg (ed.): *Recognition in Microbe-Plant Symbiotic and Pathogenic Interactions. NATO ASI Series, Vol. H4,* Springer-Verlag, Berlin, pp 253–258.

Cervone, F., De Lorenzo, G., Degrà, L., Salvi, G. and Bergami, M. (1987) Purification and characterization of a polygalacturonase-inhibiting protein from *Phaseolus vulgaris* L. *Plant Physiol.* 85: 631–637.

Cervone, F., Hahn, M.G., De Lorenzo, G., Darvill, A. and Albersheim, P. (1989) Host-pathogen interactions. XXXIII. A plant protein converts a fungal pathogenesis factor into an elicitor of plant defense responses. *Plant Physiol.* 90: 542–548.

Cervone, F., De Lorenzo, G., Caprari, C., Clark, A.J., Desiderio, A., Devoto, A., Leckie, F., Nuss, L., Salvi, G. and Toubart, P. (1993) The interaction between fungal *endo*polygalacturonase and plant cell wall PGIP (Polygalacturonase Inhibiting Protein). *In*: B. Fritig and M. Legrand (eds): *Mechanisms of Plant Defence Responses,* Kluwer Academic Publishers, Dordrecht, pp 64–67.

Chai, H.B. and Doke, N. (1987) Superoxide anion generation: a response of potato leaves to infection with *Phyto-phthora infestans*. *Phytopathology* 77: 645–649.

Chandra, S. and Low, P.S. (1995) Role of phosphorylation in elicitation of the oxidative burst in cultured soybean cells. *Proc. Natl. Acad. Sci. USA* 92: 4120–4123.

Cheong, J.-J. and Hahn, M.G. (1991) A specific, high-affinity binding site for the hepta-β-glucoside elicitor exists in soybean membranes. *Plant Cell* 3: 137–147.

Cheong, J.-J., Birberg, W., Fügedi, P., Pilotti, Å., Garegg, P.J., Hong, N., Ogawa, T. and Hahn, M.G. (1991) Structure-activity relationships of oligo-β-glucoside elicitors of phytoalexin accumulation in soybean. *Plant Cell* 3: 127–136.

Cheong, J.-J., Alba, R., Côté, F., Enkerli, J. and Hahn, M.G. (1993) Solubilization of functional plasma membrane-localized hepta-β-glucoside elicitor-binding proteins from soybean. *Plant Physiol.* 103: 1173–1182.

Choi, D. and Bostock, R.M. (1994) Involvement of the novo protein synthesis, protein kinase, extracellular Ca^{2+}, and lipoxygenase in arachidonic acid induction of 3-hydroxy-3-methylglutaryl coenzyme A reductase genes and isoprenoid accumulation in potato (*Solanum tuberosum* L.). *Plant Physiol.* 104: 1237–1244.

Cooke, C.J., Smith, C.J., Walton, T.J. and Newton, R.P. (1994) Evidence that cyclic AMP is involved in the hypersensitive response of *Medicago sativa* to a fungal elicitor. *Phytochemistry* 35: 889–895.

Cosio, E.G., Pöpperl, H., Schmidt, W.E. and Ebel, J. (1988) High-affinity binding of fungal β-glucan fragments to soybean (*Glycine max* L.) microsomal fractions and protoplasts. *Europ. J. Biochem.* 175: 309–315.

Cosio, E.G., Frey, T. and Ebel, J. (1990a) Solubilization of soybean membrane binding sites for fungal β-glucans that elicit phytoalexin accumulation. *FEBS Lett.* 264: 235–238.

Cosio, E.G., Frey, T., Verduyn, R., Van Boom, J. and Ebel, J. (1990b) High-affinity binding of a synthetic hepta-glucoside and fungal glucan phytoalexin elicitors to soybean membranes. *FEBS Lett.* 271: 223–226.

Cosio, E.G., Frey, T. and Ebel, J. (1992) Identification of a high-affinity binding protein for a hepta-β-glucoside phytoalexin elicitor in soybean. *Europ. J. Biochem.* 204: 1115–1123.

Côté, F. and Hahn, M.G. (1994) Oligosaccharins: structures and signal transduction. *Plant Molec. Biol.* 26: 1379–1411.

Côté, F., Cheong, J.-J., Alba, R. and Hahn, M.G. (1995) Characterization of binding proteins that recognize oligo-glucoside elicitors of phytoalexin synthesis in soybean. *Physiol. Plant.* 93: 401–410.

Dangl, J.L. (1995) Pièce de résistance: Novel classes of plant disease resistance genes. *Cell* 80: 363–366.

Darvill, A., Augur, C., Bergmann, C., Carlson, R.W., Cheong, J.-J., Eberhard, S., Hahn, M.G., Ló, V.-M., Marfà, V., Meyer, B., Mohnen, D., O'Neill, M.A., Spiro, M.D., van Halbeek, H., York, W.S. and Albersheim, P. (1992) Oligosaccharins – Oligosaccharides that regulate growth, development and defence responses in plants. *Glycobiology* 2: 181–198.

Davis, K.R. and Hahlbrock, K. (1987) Induction of defense responses in cultured parsley cells by plant cell wall fragments. *Plant Physiol.* 85: 1286–1290.

Davis, K.R., Darvill, A.G., Albersheim, P. and Dell, A. (1986) Host-pathogen interactions. XXIX. Oligogalact-uronides released from sodium polypectate by endopolygalacturonic acid lyase are elicitors of phytoalexins in soybean. *Plant Physiol.* 80: 568–577.

De Lorenzo, G., Ranucci, A., Bellincampi, D., Salvi, G. and Cervone, F. (1987) Elicitation of phenylalanine ammonia-lyase in *Daucus carota* by oligogalacturonides released from sodium polypectate by homogenous polygalacturonase. *Plant Sci.* 51: 147–150.

De Lorenzo, G., Ito, Y., D'Ovidio, R., Cervone, F., Albersheim, P. and Darvill, A.G. (1990) Host-pathogen interactions. XXXVII. Abilities of the polygalacturonase-inhibiting proteins from four cultivars of *Phaseolus vulgaris* to inhibit the *endo*polygalacturonases from three races of *Colletrichum lindemuthianum*. *Physiol. Molec. Plant Pathol.* 36: 421–435.

De Lorenzo, G., Cervone, F., Bellincampi, D., Caprari, C., Clark, A.J., Desiderio, A., Devoto, A., Forrest, R., Leckie, F., Nuss, L. and Salvi, G. (1994) Polygalacturonase, PGIP and oligogalacturonides in cell-cell communication. *Biochem. Soc. Trans.* 22: 396–399.

De Wit, P.J.G.M. (1995) *Cf9* and *Avr9*, two major players in the gene-for-gene game. *Trends Microbiol.* 3: 251–252.

Dean, J.F.D. and Anderson, J.D. (1991) Ethylene biosynthesis-inducing xylanase. II. Purification and physical characterization of the enzyme produced by *Trichoderma viride*. *Plant Physiol.* 95: 316–323.

Dean, J.F.D., Gross, K.C. and Anderson, J.D. (1991) Ethylene biosynthesis-inducing xylanase. III. Product characterization. *Plant Physiol.* 96: 571–576.

Diekmann, W., Herkt, B., Low, P.S., Nürnberger, T., Scheel, D., Terschüren, C. and Robinson, D.G. (1994) Visualization of elicitor-binding loci at the plant cell surface. *Planta* 195: 126–137.

Dietrich, A., Mayer, J.E. and Hahlbrock, K. (1990) Fungal elicitor triggers rapid, transient, and specific protein phosphorylation in parsley cell suspension cultures. *J. Biol. Chem.* 265: 6360–6368.

Dixon, R.A. and Lamb, C.J. (1990) Molecular communication in interactions between plants and microbial pathogens. *Annu. Rev. Plant Physiol.* 41: 339–367.

Dixon, R.A., Harrison, M.J. and Lamb, C.J. (1994) Early events in the activation of plant defense responses. *Annu. Rev. Phytopathol.* 32: 479–501.

Doares, S.H., Syrovets, T., Weiler, E.W. and Ryan, C.A. (1995) Oligogalacturonides and chitosan activate plant defensive genes through the octadecanoid pathway. *Proc. Natl. Acad. Sci. USA* 92: 4095–4098.

Ebel, J. and Cosio, E.G. (1994) Elicitors of plant defense responses. *Int. Rev. Cytol.* 148: 1–36.

Fanelli, C., Castoria, R., Fabbri, A.A. and Passi, S. (1992) Novel study on the elicitation of the hypersensitive response by polyunsaturated fatty acids in potato tuber. *Natural Toxins* 1: 136–146.

Farmer, E.E. and Helgeson, J.P. (1987) An extracellular protein from *Phytophthora parasitica* var. *nicotianae* is associated with stress metabolite accumulation in tobacco callus. *Plant Physiol.* 85: 733–740.

Farmer, E.E. and Ryan, C.A. (1992) Octadecanoid precursors of jasmonic acid activate the synthesis of wound-inducible proteinase inhibitors. *Plant Cell* 4: 129–134.

Farmer, E.E., Moloshok, T.D., Saxton, M.J. and Ryan, C.A. (1991) Oligosaccharide signaling in plants: Specificity of oligouronide-enhanced plasma membrane protein phosphorylation. *J. Biol. Chem.* 266: 3140–3145.

Favaron, F., Alghisi, P. and Marciano, P. (1992) Characterization of two *Sclerotinia sclerotiorum* polygalacturonases with different abilities to elicit glyceollin in soybean. *Plant Sci.* 83: 7–13.

Felix, G., Grosskopf, D.G., Regenass, M. and Boller, T. (1991) Rapid changes of protein phosphorylation are involved in transduction of the elicitor signal in plant cells. *Proc. Natl. Acad. Sci. USA* 88: 8831–8834.

Felix, G., Regenass, M. and Boller, T. (1993) Specific perception of subnanomolar concentrations of chitin fragments by tomato cells: Induction of extracellular alkalinization, changes in protein phosphorylation, and establishment of a refractory state. *Plant J.* 4: 307–316.

Felix, G., Regenass, M., Spanu, P. and Boller, T. (1994) The protein phosphatase inhibitor calyculin A mimics elicitor action in plant cells and induces rapid hyperphosphorylation of specific proteins as revealed by pulse labeling with [^{33}P]phosphate. *Proc. Natl. Acad. Sci. USA* 91: 952–956.

Frediani, M., Cremonini, R., Salvi, G., Caprari, C., Desiderio, A., D'Ovidio, R., Cervone, F. and De Lorenzo, G. (1993) Cytological localization of the *pgip* genes in the embryo suspensor cells of *Phaseolus vulgaris* L. *Theoret. Appl. Genet.* 87: 369–373.

Frey, T., Cosio, E.G. and Ebel, J. (1993) Affinity purification and characterization of a binding protein for a hepta-β-glucoside phytoalexin elicitor in soybean. *Phytochemistry* 32: 543–550.

Fuchs, Y., Saxena, A., Gamble, H.R. and Anderson, J.D. (1989) Ethylene biosynthesis-inducing protein from cellulysin is an endoxylanase. *Plant Physiol.* 89: 138–143.

Granado, J., Felix, G. and Boller, T. (1995) Perception of fungal sterols in plants. Subnanomolar concentrations of ergosterol elicit extracellular alkalinization in tomato cells. *Plant Physiol.* 107: 485–490.

Grosskopf, D.G., Felix, G. and Boller, T. (1990) K-252a inhibits the response of tomato cells to fungal elicitors *in vivo* and their microsomal protein kinase *in vitro*. *FEBS Lett.* 275: 177–180.

Gundlach, H., Müller, M.J., Kutchan, T.M. and Zenk, M.H. (1992) Jasmonic acid is a signal transducer in elicitor-induced plant cell cultures. *Proc. Natl. Acad. Sci. USA* 89: 2389–2393.

Hahlbrock, K., Scheel, D., Logemann, E., Nürnberger, T., Parniske, M., Reinold, S., Sacks, W.R. and Schmelzer, E. (1995) Oligopeptide elicitor-mediated defense gene activation in cultured parsley cells. *Proc. Natl. Acad. Sci. USA* 92: 4150–4157.

Hahn, M.G., Darvill, A.G. and Albersheim, P. (1981) Host-pathogen interactions. XIX. The endogenous elicitor, a fragment of a plant cell wall polysaccharide that elicits phytoalexin accumulation in soybeans. *Plant Physiol.* 68: 1161–1169.

Hahn, M.G., Bucheli, P., Cervone, F., Doares, S.H., O'Neill, R.A., Darvill, A. and Albersheim, P. (1989) Roles of cell wall constituents in plant-pathogen interactions. *In*: T. Kosuge and E.W. Nester (eds): *Plant-Microbe Interactions. Molecular and Genetic Perspectives, Vol. 3*, McGraw Hill Publishing Co. New York, NY, pp 131–181.

Harper, J.F., Sussman, M.R., Schaller, G.E., Putnam-Evans, C., Charbonneau, H. and Harmon, A.C. (1991) A calcium-dependent protein kinase with a regulatory domain similar to calmodulin. *Science* 252: 951–954.

Holden, F.R. and Walton, J.D. (1992) Xylanases from the fungal maize pathogen *Cochliobolus carbonum*. *Physiol. Molec. Plant Pathol.* 40: 39–47.

Horn, M.A., Heinstein, P.F. and Low, P.S. (1989) Receptor-mediated endocytosis in plant cells. *Plant Cell* 1: 1003–1009.

Hoshino, T., Chida, M., Yamaura, T., Yoshizawa, Y. and Mizutani, J. (1994) Phytoalexin induction in green pepper cell cultures treated with arachidonic acid. *Phytochemistry* 36: 1417–1419.

Jacinto, T., Farmer, E.E. and Ryan, C.A. (1993) Purification of potato leaf plasma membrane protein pp 34, a protein phosphorylated in response to oligogalacturonide signals for defense and development. *Plant Physiol.* 103: 1393–1397.

Jones, D.A., Thomas, C.M., Hammond-Kosack, K.E., Balint-Kurti, P.J. and Jones, J.D.G. (1994) Isolation of the tomato *Cf-9* gene for resistance to *Cladosporium fulvum* by transposon tagging. *Science* 266: 789–793.

Joosten, M.H.A.J., Cozijnsen, T.J. and de Wit, P.J.G.M. (1994) Host resistance to a fungal tomato pathogen lost by a single base-pair change in an avirulence *Gene*. *Nature* 367: 384–386.

Kamada, Y. and Muto, S. (1994) Stimulation by fungal elicitor of inositol phospholipid turnover in tobacco suspension culture cells. *Plant Cell Physiol.* 35: 397–404.

Kamoun, S., Young, M., Glascock, C.B. and Tyler, B.M. (1993) Extracellular protein elicitors from *Phytophthora*: Host-specificity and induction of resistance to bacterial and fungal phytopathogens. *Molec. Plant Microbe Interactions* 6: 15–25.

Kauss, H., Jeblick, W., Ziegler, J. and Krabler, W. (1994) Pretreatment of parsley (*Pretoselinum crispum* L.) suspension cultures with methyl jasmonate enhances elicitation of activated oxygen species. *Plant Physiol.* 105: 89–94.

Kendra, D.F. and Hadwiger, L.A. (1984) Characterization of the smallest chitosan oligomer that is maximally antifungal to *Fusarium solani* and elicits pisatin formation in *Pisum sativum. Exp. Mycol.* 8: 276–281.

Knight, M.R., Campbell, A.K., Smith, S.M. and Trewavas, A.J. (1991) Transgenic plant aequorin reports the effects of touch and cold-shock and elicitors on cytoplasmic calcium. *Nature* 352: 524–526.

Kobe, B. and Deisenhofer, J. (1994) The leucine-rich repeat: A versatile binding motif. *Trends Biochem. Sci.* 19: 415–421.

Kobe, B. and Deisenhofer, J. (1995) A structural basis of the interactions between leucine-rich repeats and protein ligands. *Nature* 374: 183–186.

Kogel, G., Beissmann, B., Reisener, H.J. and Kogel, K. (1991) Specific binding of a hypersensitive lignification elicitor from *Puccinia graminis* f. sp. *tritici* to the plasma membrane from wheat (*Triticum aestivum* L.). *Planta* 183: 164–169.

Kohn, R. (1975) Ion binding on polyuronates-alginate and pectin. *Pure Appl. Chem.* 42: 371–397.

Kurosaki, F., Tsurusawa, Y. and Nishi, A. (1987) The elicitation of phytoalexins by Ca^{2+} and cyclic AMP in carrot cells. *Phytochemistry* 26: 1919–1923.

Lafitte, C., Barthe, J.-P., Gansel, X., Dechamp-Guillaume, G., Faucher, C., Mazau, D. and Esquerré-Tugayé, M.-T. (1993) Differential induction by endopolygalacturonase of β-1,3-glucanases in *Phaseolus vulgaris* isoline susceptible and resistant to *Colletotrichum lindemuthianum* race β. *Molec. Plant Microbe Interactions* 6: 628–634.

Lawrence, G.J., Finnegan, E.J., Ayliffe, M.A. and Ellis, J.G. (1995) The *L6* gene for flax rust resistance is related to the *Arabidopsis* bacterial resistance gene *RPS2* and the tobacco viral resistance gene *N. Plant Cell* 7: 1195–1206.

Lee, S.C. and West, C.A. (1981) Properties of *Rhizopus stolonifer* polygalacturonase, an elicitor of casbene synthetase activity in castor bean (*Rhizopus stolonifer* L.) seedlings. *Plant Physiol.* 67: 640–645.

Legendre, L., Yueh, Y.G., Crain, R.C., Haddock, N., Heinstein, P.F. and Low, P.S. (1993) Phospholipase C activation during elicitation of the oxidative burst in cultured plant cells. *J. Biol. Chem.* 268: 24559–24563.

Levine, A., Tenhaken, R., Dixon, R. and Lamb, C. (1994) H_2O_2 from the oxidative burst orchestrates the plant hypersensitive disease response. *Cell* 79: 583–593.

Liners, F., Letesson, J.-J., Didembourg, C. and Van Cutsem, P. (1989) Monoclonal antibodies against pectin. Recognition of a conformation induced by calcium. *Plant Physiol.* 91: 1419–1424.

Longland, A.C., Slusarenko, A.J. and Friend, J. (1987) Arachidonic and linoleic acids elicit isoflavonoid phytoalexin accumulation in *Phaseolus vulgaris* (French bean). *J. Phytopathol.* 120: 289–297.

Lotan, T. and Fluhr, R. (1990) Xylanase, a novel elicitor of pathogenesis-related proteins in tobacco, uses a non-ethylene pathway for induction. *Plant Physiol.* 93: 811–817.

Marinelli, F., Di Gregorio, S. and Nuti Ronchi, V. (1991) Phytoalexin production and cell death in elicited carrot cell suspension cultures. *Plant Sci.* 77: 261–266.

Martin, G.B., Brommonschenkel, S.H., Chunwongse, J., Frary, A., Ganal, M.W., Spivey, R., Wu, T., Earle, E.D. and Tanksley, S.D. (1993) Map-based cloning of a protein kinase gene conferring disease resistance in tomato. *Science* 262: 1432–1436.

Mathieu, Y., Kurkdjian, A., Xia, H., Guern, J., Koller, A., Spiro, M., O'Neill, M., Albersheim, P. and Darvill, A. (1991) Membrane responses induced by oligogalacturonides in suspension-cultured tobacco cells. *Plant J.* 1: 333–343.

Messiaen, J. and Van Cutsem, P. (1993) Defense gene transcription in carrot cells treated with oligogalacturonides. *Plant Cell Physiol.* 34: 1117–1123.

Messiaen, J. and Van Cutsem, P. (1994) Pectic signal transduction in carrot cells: Membrane, cytosolic and nuclear responses induced by oligogalacturonides. *Plant Cell Physiol.* 35: 677–689.

Messiaen, J., Read, N.D., Van Cutsem, P. and Trewavas, A.J. (1993) Cell wall oligogalacturonides increase cytosolic free calcium in carrot protoplasts. *J. Cell Sci.* 104: 365–371.

Mueller, M.J., Brodschelm, W., Spannagl, E. and Zenk, M.H. (1993) Signaling in the elicitation process is mediated through the octadecanoid pathway leading to jasmonic acid. *Proc. Natl. Acad. Sci. USA* 90: 7490–7494.

Nasrallah, J.B. and Nasrallah, M.E. (1993) Pollen-stigma signaling in the sporophytic self-incompatibility response. *Plant Cell* 5: 1325–1335.

Nasrallah, J.B., Stein, J.C., Kandasamy, M.K. and Nasrallah, M.E. (1994) Signaling the arrest of pollen tube development in self-incompatible plants. *Science* 266: 1505–1508.

Nespoulous, C., Huet, J.-C. and Pernollet, J.-C. (1992) Structure-function relationships of α and β elicitins, signal proteins involved in the plant-*Phytophthora* interaction. *Planta* 186: 551–557.

Nuss, L., Mahé, A., Clark, A.J., Grisvard, J., Dron, M., Cervone, F. and De Lorenzo, G. (1996) Differential accumulation of polygalacturonase-inhibiting protein (PGIP) mRNA in two near-isogenic lines of *Phaseolus vulgaris* L. upon infection with *Colletotrichum lindemuthianum. Physiol. Molec. Plant Pathol.* 48: 83–89.

Nürnberger, T., Nennstiel, D., Jabs, T., Sacks, W.R., Hahlbrock, K. and Scheel, D. (1994) High affinity binding of a fungal oligopeptide elicitor to parsley plasma membranes triggers multiple defense responses. *Cell* 78: 449–460.

Nürnberger, T., Nennstiel, D., Hahlbrock, K. and Scheel, D. (1995) Covalent cross-linking of the *Phytophthora megasperma* oligopeptide elicitor to its receptor in parsley membranes. *Proc. Natl. Acad. Sci. USA* 92: 2338–2342.

Pernollet, J.C., Sallantin, M., Sallé-Tourne, M. and Huet, J.C. (1993) Elicitin isoforms from seven *Phytophthora* species: comparison of their physico-chemical properties and toxicity to tobacco and other plant species. *Physiol. Molec. Plant Pathol.* 42: 53–67.

Pouteau, S., Grandbastien, M.-A. and Boccara, M. (1994) Microbial elicitors of plant defence responses activate transcription of a retrotransposon. *Plant J.* 5: 535–542.

Preisig, C.L. and Kuc, J. (1985) Arachidonic acid-related elicitors of the hypersensitive response in potato and enhancement of their activities by glucans from *Phytophthora infestans* (Mont.) de Bary. *Arch. Biochem. Biophys.* 236: 379–389.

Ren, Y.-Y. and West, C.A. (1992) Elicitation of diterpene biosynthesis in rice (*Oryza sativa* L.) by chitin. *Plant Physiol.* 99: 1169–1178.

Renelt, A., Colling, C., Hahlbrock, K., Nürnberger, T., Parker, J.E., Sacks, W.R. and Scheel, D. (1993) Studies on elicitor recognition and signal transduction in plant defence. *J. Exp. Bot.* 44 Suppl. 257–268.

Reymond, P., Grünberger, S., Paul, K., Müller, M. and Farmer, E.E. (1995) Oligogalacturonide defense signals in plants: Large fragments interact with the plasma membrane *in vitro*. *Proc. Natl. Acad. Sci. USA* 92: 4145–4149.

Ricker, K.E. and Bostock, R.M. (1994) Eicosanoids in the *Phytophthora infestans*-potato interaction: Lipoxygenase metabolism of arachidonic acid and biological activities of selected lipoxygenase products. *Physiol. Molec. Plant Pathol.* 44: 65–80.

Robertsen, B. (1986) Elicitors of the production of lignin-like compounds in cucumber hypocotyls. *Physiol. Molec. Plant Pathol.* 28: 137–148.

Robertsen, B. (1987) Endo-polygalacturonase from *Cladosporium cucumerinum* elicits lignification in cucumber hypocotyls. *Physiol. Molec. Plant Pathol.* 31: 361–374.

Roby, D., Toppan, A. and Esquerré-Tugayé, M.-T. (1987) Cell surfaces in plant micro-organism interactions. VIII. Increased proteinase inhibitor activity in melon plants in response to infection by *Colletotrichum lagenarium* or to treatment with an elicitor fraction from this fungus. *Physiol. Molec. Plant Pathol.* 30: 453–460.

Rogers, K.R., Albert, F. and Anderson, A.J. (1988) Lipid peroxidation is a consequence of elicitor activity. *Plant Physiol.* 86: 547–553.

Rohe, M., Gierlich, A., Hermann, H., Hahn, M., Schmidt, B., Rosahl, S. and Knogge, W. (1995) The race-specific elicitor, NIP1, from the barley pathogen, *Rhynchosporium secalis*, determines avirulence on host plants of the *Rrs1* resistance genotype. *EMBO J.* 14: 4168–4177.

Schmidt, W.E. and Ebel, J. (1987) Specific binding of a fungal glucan phytoalexin elicitor to membrane fractions from soybean *Glicine max*. *Proc. Natl. Acad. Sci. USA* 84: 4117–4121.

Schreck, R., Rieber, P. and Bauerle, P.A. (1991) Reactive oxygen intermediates are apparently widely used messengers in the activation of the NF-kB transcription factor and HIV-1. *EMBO J.* 10: 2247–2258.

Schwacke, R. and Hager, A. (1992) Fungal elicitors induce a transient release of active oxygen species from cultured spruce cells that is dependent on Ca^{2+} and protein-kinase activity. *Planta* 187: 136–141.

Sharon, A., Fuchs, Y. and Anderson, J.D. (1993) The elicitation of ethylene biosynthesis by a *Trichoderma* xylanase is not related to the cell wall degradation activity of the enzyme. *Plant Physiol.* 102: 1325–1329.

Sharp, J.K., Valent, B. and Albersheim, P. (1984) Purification and partial characterization of a β-glucan fragment that elicits phytoalexin accumulation in soybean. *J. Biol. Chem.* 259: 11312–11320.

Shibuya, N., Kaku, H., Kuchitsu, K. and Maliarik, M.J. (1993) Identification of a novel high-affinity binding site for *N*-acetylchitooligosaccharide elicitor in the membrane fraction from suspension-cultured rice cells. *FEBS Lett.* 329: 75–78.

Song, W.Y., Wang, G.L., Chen, L.L., Kim, H.S., Pi, L.Y., Holsten, T., Gardner, J., Wang, B., Zhai, W.X., Zhu, L.H., Fauquet, C. and Ronald, P. (1995) A receptor kinase-like protein encoded by the rice disease resistance gene, *Xa21*. *Science* 270: 1804–1806.

Staehelin, C., Granado, J., Müller, J., Wiemken, A., Mellor, R.B., Felix, G., Regenass, M., Broughton, W.J. and Boller, T. (1994) Perception of *Rhizobium* nodulation factors by tomato cells and inactivation by root chitinases. *Proc. Natl. Acad. Sci. USA* 91: 2196–2200.

Staskawicz, B.J., Ausubel, F.M., Baker, B.J., Ellis, J.G. and Jones, J.D.G. (1995) Molecular genetics of plant disease resistance. *Science* 268: 661–667.

Stäb, M.R. and Ebel, J. (1987) Effects of Ca^{2+} on phytoalexin induction by fungal elicitor in soybean cells. *Arch. Biochem. Biophys.* 257: 416–423.

Suzuki, K. and Shinshi, H. (1995) Transient activation and tyrosine phosphorylation of a protein kinase in tobacco cells treated with a fungal elicitor. *Plant Cell* 7: 639–647.

Svalheim, O. and Robertsen, B. (1993) Elicitation of H_2O_2 production in cucumber hypocotyl segments by oligo-1,4-α-D-galacturonides and an oligo-β-glucan preparation from cell walls of *Phytophthora megasperma* f.sp. *glycinea*. *Physiol. Plant.* 88: 675–681.

Templeton, M.D., Rikkerink, E.H.A. and Beever, R.E. (1994) Small, cysteine-rich proteins and recognition in fungal-plant interactions. *Molec. Plant Microbe Interactions* 7: 320–325.

Tenhaken, R., Levine, A., Brisson, L.F., Dixon, R.A. and Lamb, C. (1995) Function of the oxidative burst in hypersensitive disease resistance. *Proc. Natl. Acad. Sci. USA* 92: 4158–4163.

Thain, J.F., Gubb, I.R. and Wildon, D.C. (1995) Depolarization of tomato leaf cells by oligogalacturonide elicitors. *Plant Cell Environ.* 18: 211–214.

Toubart, P., Desiderio, A., Salvi, G., Cervone, F., Daroda, L., De Lorenzo, G., Bergmann, C., Darvill, A.G. and Albersheim, P. (1992) Cloning and characterization of the gene encoding the endopolygalacturonase- inhibiting protein (PGIP) of *Phaseolus vulgaris* L. *Plant J.* 2: 367–373.

Toyoda, K., Shiraishi, T., Yamada, T., Ichinose, Y. and Oku, H. (1993) Rapid changes in polyphosphoinositide metabolism in pea in response to fungal signals. *Plant Cell Physiol.* 34: 729–735.

van den Ackerveken, G.F.J.M., van Kan, J.A.L. and de Wit, P.J.G.M. (1992) Molecular analysis of the avirulence gene *avr9* of the fungal tomato pathogen *Cladosporium fulvum* fully supports the gene-for-gene hypothesis. *Plant J.* 2: 359–366.

van den Ackerveken, G.F.J.M., Vossen, P. and de Wit, P.J.G.M. (1993) The AVR9 race-specific elicitor of *Cladosporium fulvum* is processed by endogenous and plant proteases. *Plant Physiol.* 103: 91–96.

Vaughn, S.F. and Lulai, E.C. (1992) Further evidence that lipoxygenase activity is required for arachidonic acid-elicited hypersensitivity in potato callus cultures. *Plant Sci.* 84: 91–98.

Vera-Estrella, R., Barkla, B.J., Higgins, V.J. and Blumwald, E. (1994a) Plant defense response to fungal pathogens. Activation of host-plasma membrane H+-ATPase by elicitor-induced enzyme dephosphorylation. *Plant Physiol.* 104: 209–215.

Vera-Estrella, R., Higgins, V.J. and Blumwald, E. (1994b) Plant defense response to fungal pathogens. II. G-protein-mediated changes in host plasma membrane redox reactions. *Plant Physiol.* 106: 97–102.

Viard, M.P., Martin, F., Pugin, A., Ricci, P. and Blein, J.P. (1994) Protein phosphorylation is induced in tobacco cells by the elicitor cryptogein. *Plant Physiol.* 104: 1245–1249.

Vögeli, U., Vögeli-Lange, R. and Chappell, J. (1992) Inhibition of phytoalexin biosynthesis in elicitor-treated tobacco cell-suspension cultures by calcium/calmodulin antagonists. *Plant Physiol.* 100: 1369–1376.

Walker, J.C. (1993) Receptor-like protein kinase genes of *Arabidopsis thaliana*. *Plant J.* 3: 451–456.

Walker-Simmons, M., Hadwiger, L. and Ryan, C.A. (1983) Chitosans and pectic polysaccharides both induce the accumulation of the antifungal phytoalexin pisatin in pea pods and antinutrient proteinase inhibitors in tomato leaves. *Biochem. Biophys. Res. Comm.* 110: 194–199.

Walker-Simmons, M., Jin, D., West, C.A., Hadwiger, L. and Ryan, C.A. (1984) Comparison of proteinase inhibitor-inducing activities and phytoalexin elicitor activities of a pure fungal endopolygalacturonase, pectic fragments, and chitosan. *Plant Physiol.* 76: 833–836.

Wendehenne, D., Binet, M.N., Blein, J.P., Ricci, P. and Pugin, A. (1995) Evidence for specific, high-affinity binding sites for a proteinaceous elicitor in tobacco plasma membrane. *FEBS Lett.* 374: 203–207.

Yoshikawa, M., Keen, N.T. and Wang, M.-C. (1983) A receptor on soybean membranes for a fungal elicitor of phytoalexin accumulation. *Plant Physiol.* 73: 497–506.

Yu, L.M. (1995) Elicitins from *Phytophthora* and basic resistance in tobacco. *Proc. Natl. Acad. Sci. USA* 92: 4088–4094.

Zhang, S. and Mehdy, M.C. (1994) Binding of a 50-kD protein to a U-rich sequence in an mRNA encoding a proline-rich protein that is destabilized by fungal elicitor. *Plant Cell* 6: 135–145.

Zhang, S., Sheng, J., Liu, Y. and Mehdy, M.C. (1993) Fungal elicitor-induced bean proline-rich protein mRNA down-regulation is due to destabilization that is transcription and translation dependent. *Plant Cell* 5: 1089–1099.

Subject index

B. Sobral
California Institute of Biological Research (CIBR), La Jolla, CA, USA (Ed.)

The Impact of Plant Molecular Genetics

1996. 364 pages. Hardcover.
ISBN 3-7643-3802-4

Recent years have witnessed an explosion of molecular genetic techniques and approaches that have enabled large gains in basic and applied plant genetics. These techniques span the realm of genetic transformation, which is the non-sexual introduction of new genes in plants, to DNA-marker-assisted genetic studies. DNA markers, in particular, have allowed previously intractable problems, such as genetics of polypoinds and phylogenetics and evolution of crops, to be studied in great detail. DNA markers are also largely responsible for bridging qualitative, quantitative and developmental genetics, thereby providing a rich, multidisciplinary environment for plant biology. The application of DNA markers and plant transformation technologies have serious socio-economic implications for world agriculture, especially in the developing world.

THE IMPACT OF PLANT MOLECULAR GENETICS

Bruno W.S. Sobral
Editor

BIRKHÄUSER

The chapters in this volume, written by international experts in diverse fields, provide not only an update of state of the art techniques in many crucial areas of plant genetics, but also look forward into the future by defining current bottlenecks and research goals. Special attention has been given to DNA markers and their applications, but a section on social and economic implications integrates laboratory science with the socio-economic realities in which they occur.

Scientists and students in both the biological and social sciences will learn a lot from this book about the recent advances and future directions of plant genetics. In addition, this book should be required reading for policymakers.

Birkhäuser Verlag • Basel • Boston • Berlin

S. Papa, *Institute of Medical Biochemistry and Chemistry, University of Bari, Italy*
J.M. Tager, *E.C. Slater Institute, University of Amsterdam, The Netherlands (Eds)*

Biochemistry
of Cell Membranes
A Compendium of Selected Topics

1995. 376 pages. Hardcover. ISBN 3-7643-5056-3 (MCBU)

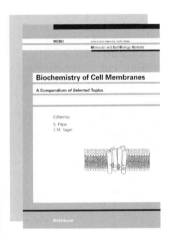

This book consists of a series of reviews on selected topics within the rapidly and vastly expanding field of membrane biology. Its aim is to highlight the most significant and important advances that have been made in recent years in understanding the structure, dynamics and functions of cell membranes.

Areas covered in this monograph include

- Signal Transduction
- Membrane Traffic: Protein and Lipids
- Bioenergetics: Energy Transfer and Membrane Transport
- Cellular Ion Homeostasis
- Growth Factors and Adhesion Molecules
- Structural Analysis of Membrane Proteins
- Membranes and Disease

Biochemistry of Cell Membranes should serve as a benchmark for indicating the most important lines for future research in these areas.

Birkhäuser Verlag • Basel • Boston • Berlin